W9-AOD-110

NONLINEAR EVOLUTION EQUATIONS

Publication No. 40
of the Mathematics Research Center
The University of Wisconsin–Madison

ACADEMIC PRESS RAPID MANUSCRIPT REPRODUCTION

NONLINEAR EVOLUTION EQUATIONS

Edited by Michael G. Crandall

Proceedings of a Symposium
Conducted by the Mathematics Research Center
The University of Wisconsin–Madison
October 17–19, 1977

ACADEMIC PRESS
New York San Francisco London 1978
A Subsidiary of Harcourt Brace Jovanovich, Publishers

This work was sponsored by the United States Army under Contract No. DAAG29-75-C-0024, the National Science Foundation under Grant No. MCS77-07583, and related to the Department of the Navy Research Grant No. N00014-77-G-0063 issued by the Office of Naval Research. The United States Government has a royalty-free license throughout the world for its own use in all copyrightable material contained herein.

ACADEMIC PRESS, INC.
111 Fifth Avenue, New York, New York 10003

United Kingdom Edition published by
ACADEMIC PRESS, INC. (LONDON) LTD.
24/28 Oval Road, London NW1 7DX

Library of Congress Cataloging in Publication Data

Symposium on Nonlinear Evolution Equations, University of Wisconsin-Madison, 1977.
Nonlinear evolution equations.

(Publication of the Mathematics Research Center, the University of Wisconsin-Madison ; no. 40)
"Contract no. DAAG29-75-0-0024".

Includes index.
1. Differential equations, Nonlinear—Congresses.
2. Differential equations, Partial—Congresses.
I. Crandall, Michael G. II. Wisconsin. University-Madison. Mathematics Research Center. III. Title.
IV. Title: Evolution equations. V. Series: Wisconsin. University-Madison. Mathematics Research Center.
QA3. U45 no. 40 [QA370] 510'.8s [515'.35] 78-12744
ISBN 0-12-195250-9

PRINTED IN THE UNITED STATES OF AMERICA

Contents

List of Contributors

Numbers in parentheses indicate the pages on which authors' contributions begin.

Invited Speakers

J. M. Ball (189), Department of Mathematics, Heriot-Watt University, Edinburgh

Haim Brezis (141), Department of Mathématiques, Université Paris VI, 75230 Paris

Alexandre Joel Chorin (17), Department of Mathematics, University of California, Berkeley, California 94720

C. M. Dafermos (103), Department of Applied Mathematics, Brown University, Providence, Rhode Island 02912

R. J. DiPerna (1), Department of Mathematics, University of Wisconsin–Madison, Madison, Wisconsin 53706

Lawrence C. Evans (163), Department of Mathematics, University of Kentucky, Lexington, Kentucky 40506

Paul C. Fife (125), Department of Mathematics and Program in Applied Mathematics, University of Arizona, Tucson, Arizona 85721

Tosio Kato (155), Department of Mathematics, University of California, Berkeley, California 94720

Peter D. Lax (207), Courant Institute of Mathematical Sciences, New York University, New York 10012

J. L. Lions (59), College de France, Paris

Jürgen Moser, Courant Institute of Mathematical Sciences, New York University, New York, New York 10012

Takaaki Nishida (29), Department of Applied Mathematics and Physics, Kyoto University, Kyoto 606, Japan

Paul H. Rabinowitz (225), Department of Mathematics, University of Wisconsin–Madison, Madison, Wisconsin 53706

Walter A. Strauss (85), Department of Mathematics, Brown University, Providence, Rhode Island 02912

Session Chairmen

Charles C. Conley, University of Wisconsin–Madison

Louis Nirenberg, New York University

Amnon Pazy, Hebrew University

Roger Temam, Université de Paris XI

Preface

This volume constitutes the proceedings of the Symposium on Nonlinear Evolution Equations held in Madison, October 17–19, 1977. The thirteen papers presented here follow the order of the corresponding lectures. J. Moser's lecture, "Differential Equations Connected with Iso-Spectral Deformations," is not represented in this volume. However, the relevant material is forthcoming in a joint article by J. Moser and M. Adler in *Comm. Math. Phys.* (1978). This symposium was sponsored by the Army Research Office, the National Science Foundation, and the Office of Naval Research.

Among those who contributed to the success of the symposium John A. Nohel, our cochairman, deserves special mention. Hearty thanks are also due to the speakers, the chairmen of the sessions, our peerless symposium secretary Mrs. Gladys G. Moran, and the outstanding group of participants who made the symposium such an exceptional event. The cheerful competence of Judith Siesen was indispensable in preparing this volume.

Entropy and the Uniqueness of Solutions to Hyperbolic Conservation Laws

R. J. DiPerna

In this talk we shall discuss some results on the uniqueness of solutions to systems of conservation laws of the form

$$u_t + f(u)_x = 0, \quad -\infty < x < \infty \quad , \tag{1}$$

where $u = u(x,t) \in R^n$ and f is a smooth nonlinear mapping from R^n to R^n. We shall assume throughout that (1) is strictly hyperbolic, i.e. that the Jacobian ∇f of f has n real and distinct eigenvalues:

$$\lambda_1(u) < \cdots < \lambda_n(u) \quad .$$

Systems of this form arise in continuum mechanics. The equations of inviscid fluid dynamics, for example, form a system of three equations; the components of u represent densities of mass, momentum and total energy while the equations express the physical laws of the conservation of the corresponding three quantities. Other examples are provided by the equations of shallow water waves, magneto-hydrodynamics and in certain special cases, elasticity.

It is well known that the Cauchy problem for (1) does not have in general globally defined smooth solutions. The nonlinear structure of the eigenvalues leads to the development of discontinuities in the solution. On the other hand

ISBN 0-12-195250-9

in the broader class of weak solutions uniqueness is lost. Indeed it is possible for infinitely many weak solutions to share the same initial data. Thus the problem arises of identifying the class of stable weak solutions. For this purpose several criteria have been introduced. Before discussing the relevant criteria we shall briefly recall certain facts concerning the existence and singularities of solutions and the appropriate function spaces for (1). For results concerning the uniqueness of solutions to scalar conservation laws we refer the reader to the work of Oleinik [28,30], Vol'pert [33], Keyfitz [18] and Kruzkov [19].

One of the natural function spaces for conservation laws is the space $L^\infty \cap BV$. Here BV denotes the space of functions which have locally bounded total variation in the sense of Cesari. We recall that a real valued function $v = v(y)$, $y = (y_1, y_2, \ldots, y_k)$, defined on a region $\Omega \subset R^k$ is an element of $BV(\Omega)$ if it is locally integrable and if its first order partial derivatives are locally finite Borel measures:

$$\int v \, \partial\phi/\partial y_j \, dy = - \int \phi \, d\mu_j, \quad j = 1,2,\ldots,k \quad , \tag{2}$$

for all $\phi \in C_o^\infty(\Omega)$ where μ_j are Borel measures on Ω which satisfy

$$|\mu_j(\Omega')| < \infty \tag{3}$$

for all compact subsets $\Omega' \subset \Omega$. More generally, a function defined on $\Omega \subset R^k$ with values in R^n is an element of $BV(\Omega)$ if each of its components satisfies (2) and (3). By a <u>weak solution</u> we shall mean an element of $L^\infty \cap BV(\Omega)$, $\Omega \subset R^2$, which satisfies (1) in the sense of distributions. Thus a function u in $L^\infty \cap BV(\Omega)$ is a weak solution of (1) if and only if the measure

$$u_t + f(u)_x$$

vanishes on all Borel subsets of Ω.

This notion of weak solution generalizes the classical notion of piecewise smooth solution. For comparison we shall recall the structure of an arbitrary BV function. Let H_k

denote k-dimensional Hausdorff measure. It is known classically that if $v \in BV(\Omega)$, $\Omega \subset R^1$, then after a modification on a set with H_1 measure zero, v has the following properties. At each point of Ω, v is either continuous or has a jump discontinuity; the set of jump points is at most countable. More generally, if $v \in BV(\Omega)$, $\Omega \subset R^k$, then after a modification on a set with H_k measure zero, v has the following properties [8,33]. At each point of Ω with the possible exception of a set with H_{k-1} measure zero, v is either approximately continuous or has an approximate jump discontinuity in the Lebesgue sense; the set of jump points $\Gamma = \Gamma(v)$ is an at most countable union rectifiable sets:

$$\Gamma = \bigcup_{m=1}^{\infty} \Gamma_m \cup \tilde{\Gamma}$$

where $H_{k-1}(\tilde{\Gamma}) = 0$ and Γ_m is a compact subset of a Lipschitz "surface" of dimension $k - 1$, i.e.

$$\Gamma_m = \phi_m(K_m)$$

where $\phi_m \in \mathrm{Lip}(R^{k-1}, R^k)$ and K_m is a compact subset of R^{k-1}. At each point P of Γ there exists a well-defined unit normal ν in the sense of Federer that specifies the direction in which v experiences a jump discontinuity: v has distinct approximate limits at P in the directions $\pm\nu$ denoted by

$$\ell_{\nu} u(P) \quad \text{and} \quad \ell_{-\nu} u(P) \quad .$$

In the context of weak solutions, the set $\Gamma(u)$ is referred to as the <u>shock set</u> of the solution u. We shall refer to any Borel subset of $\Gamma(u)$ as a <u>wave of discontinuity</u> in u.

The space $L^{\infty} \cap BV$ is natural from the point of view of the existence theory for conservation laws. The Glimm difference scheme generates globally defined solutions to the Cauchy problem which lie in $L^{\infty} \cap BV$. Glimm [10] established convergence of the difference approximations for general systems of n equations in the case of initial data with small total variation. Convergence of the Glimm approximations has also been established for special classes of systems with initial data having large total variation [1,5,6,15,24,25,26,27]. In each of these cases it has been shown that the solution operator of the Glimm scheme is bounded in the total variation norm: if $TVu(\cdot,0) < \infty$ then

$$TVu(\cdot,t) \leq \text{const.} \quad , \tag{4}$$

where the constant depends only on f and $TVu(\cdot,0)$. One may regard the total variation of the solution in x at time t as representing the total magnitude of all waves in the solution at time t. Thus the total magnitude of waves is bounded uniformly in time if it is finite initially. It follows from (4) and the Riesz representation theorem for C_o^* that the distribution u_x is a locally bounded Borel measure. Since f is smooth and $u \in L^\infty$, the same is true for the distribution $f(u)_x$. We conclude from the form of the equations that u_t is a locally bounded Borel measure and that $u \in L^\infty \cap BV$.

Let us now consider the problem of characterizing the stable weak solutions to systems (1). Suppose that D is an open domain in R^n. We recall that a function $\eta: D \rightarrow R$ is an <u>entropy</u> for (1) with <u>entropy flux</u> $q: D \rightarrow R$ if all smooth solutions with range in D satisfy an additional conservation law of the form

$$\eta(u)_t + q(u)_x = 0 \quad . \tag{5}$$

For the equations of inviscid fluid dynamics, the classical entropy density serves as η and the classical entropy flux density serves as q; equation (5) expresses the fact that during the smooth flow of a fluid the entropy of each fluid particle remains constant in time. In general the existence of the pair (η,q) is based on a set of n compatibility conditions. If u is a smooth solution then by the chain rule

$$0 = \eta_t + q_x = \nabla\eta\, u_t + \nabla q\, u_x = (-\nabla\eta\nabla f + \nabla q)u_x \quad .$$

Thus the condition,

$$\nabla\eta\nabla f = \nabla q \quad , \tag{6}$$

is necessary and sufficient for the existence of an entropy-entropy flux pair. Equation (6) is a system of n equations in 2 unknowns. It is possible to work with only one unknown η by replacing (6) with the second order system

$$\text{curl}(\nabla\eta\nabla f) = 0 \quad , \tag{7}$$

in which case q is uniquely determined by η up to an additive constant. Although (6) and (7) are overdetermined if $n > 2$, Friedrichs and Lax [9] have observed that most of the conservative systems which result from continuum mechanics are endowed with a globally defined strictly convex entropy η, e.g. the equations of inviscid fluid dynamics and magneto-fluid dynamics, the equations of shallow water waves, general symmetric hyperbolic systems and in certain special cases the equations of elasticity, cf. [7] for a list of the corresponding entropies. In the case $n = 2$, Lax [21] has shown how to construct locally defined strictly convex entropies for general systems (1) and globally defined strictly convex entropies for a very broad class of systems (1).

Henceforth, we shall restrict our attention to those systems (1) which possess a strictly convex entropy η. For such systems Lax [21] and Kruzkov [19] have proposed the following <u>entropy criterion</u>: a weak solution u is <u>admissible</u> if

$$\theta_u \equiv \eta(u)_t + q(u)_x$$

is non-positive. (Here the range of u is assumed to be contained within the domain of definition of η and q.) We shall refer to θ_u as the <u>dissipation measure</u> for u. It is not difficult to prove that the dissipation measure is concentrated on the shock set:

$$\theta_u(B) = \theta_u(B \cap \Gamma(u)) \tag{8}$$

for all Borel sets B. In view of (8), we shall call a wave of discontinuity admissible if the restriction of θ_u to E is non-positive, i.e. if

$$\theta_u(B) \leq 0$$

for all Borel sets $B \subset E$. Thus, a weak solution is admissible if and only if all of its waves of discontinuity are admissible.

Using (8), the entropy criterion may be formulated in terms of the rate at which waves dissipate entropy as follows. Let $\sigma = \sigma(P)$ denote the speed of propagation of a wave of discontinuity E at the point P:

$$\sigma = -\nu_t(P)/\nu_x(P)$$

where $\nu(P) = (\nu_x(P), \nu_t(P))$ is the unit normal to E at P. For convenience let us normalize ν so that $\nu_x < 0$. (Since $u \in L^\infty$, $\nu_x \neq 0$ at H_1 almost all points of Γ. In the case where E is a smooth curve given by $(x(t),t)$, σ equals the classical speed of propagation $x'(t)$). It follows from the generalized Green's theorem for measures [8,33] that

$$\theta_u(E) = \int_E \nu_t[\eta] + \nu_x[q] \, dH_1$$

where bracket denotes the jump in the enclosed quantity across E:

$$[\eta] = \eta(\ell_\nu u) - \eta(\ell_{-\nu} u)$$

$$[q] = q(\ell_\nu u) - q(\ell_{-\nu} u) \quad .$$

Thus, the rate at which the wave E dissipates entropy at the point P is given by the quantity

$$\sigma[\eta] - [q]$$

evaluated at P; indeed

$$\theta_u(E) = \int_E \sigma[\eta] - [q] \, dt$$

where the measure dt denotes the restriction of $-\nu_x dH_1$ to $\Gamma(u)$. It follows that a wave of discontinuity E is admissible if and only if

$$\sigma[\eta] - [q] \leq 0 \tag{9}$$

at H_1 almost all points of E. The local formulation (9) is due to Lax [21]. Using (9) Lax [21] has shown that solutions constructed by the Glimm difference scheme and the Lax-Friedrichs scheme necessarily satisfy the entropy criterion.

We recall that the entropy criterion has been put forth in order to characterize the stable solutions to a specific class of systems (1), the class of systems with the property that each eigenvalue is either genuinely nonlinear or linearly degenerate in the sense of Lax [20], i.e. either

$$r_j \cdot \nabla\lambda_j \neq 0 \quad \text{or} \quad r_j \cdot \nabla\lambda_j \equiv 0 \tag{10}$$

where $r_j = r_j(u)$ denotes the right eigenvector of ∇f corresponding to λ_j. We note that this class of systems includes the aforementioned equations of shallow water waves, fluid dynamics, magneto-fluid dynamics and in certain cases elasticity. For systems satisfying (10) Lax [21] has shown that the entropy criterion is equivalent to the Lax shock criterion [20] (at least for solutions with moderate oscillation) which govern the number and type of characteristic which impinge on a wave of discontinuity. At the level of linearized stability the Lax shock criterion is necessary and sufficient for the stability of the solution. For systems with general eigenvalues, i.e. eigenvalues which do not satisfy (10), it is known that the entropy criterion is not sufficiently strong to rule out all unstable solutions and a more powerful criterion is needed. We refer the reader to the work of Oleinik [28], Wendroff [34,35], Dafermos [2,3] and Liu [22,23] concerning admissibility criteria for equations with general eigenvalues. Henceforth, we shall restrict our attention to the class of systems (1) with a strictly convex entropy and eigenvalues satisfying (10).

We shall conclude our discussion of the entropy criterion by recalling its thermodynamic interpretation in the setting of fluid dynamics. For this purpose, we first note that from the point of view of dissipation distinguished roles are played by waves which propagate at characteristic speed and those which do not. By a <u>contact discontinuity</u> we shall mean a wave of discontinuity E such that for some j,

$$\sigma(P) = \lambda_j\{\ell_\nu u(P)\} = \lambda_j\{\ell_{-\nu} u(P)\} \tag{11}$$

at H_1 almost all points P of E. We shall refer to any wave of discontinuity which is not a contact discontinuity as a <u>shock wave</u>. It is not difficult to prove that a wave of discontinuity is a contact discontinuity only if the corresponding eigenvalue is linearly degenerate and consequently that contact discontinuities do not dissipate entropy; i.e.

$$\theta_u(B) = 0$$

for all Borel subsets B of a contact discontinuity E. Thus, in general the entropy criterion places a restriction only on the shock waves of a solution. In the particular case of the equations of inviscid fluid dynamics the eigenvalues λ_1 and λ_3 are genuinely nonlinear while λ_2 is linearly degenerate and equal to the fluid velocity. Thus, the fluid velocity is continuous across contact discontinuities and it follows from (11) that fluid particles do not cross contact discontinuities. The entropy criterion expresses the second law of thermodynamics by requiring that the entropy of a fluid particle increase upon crossing a shock wave.

Consider a system (1) with a strictly convex entropy and eigenvalues satisfying (10). For such a system the uniqueness problem may be formulated as follows. Let

$$\mathcal{S}(T) = \{(x,t) : 0 \leq t < T\}$$

and let $K = K\{\mathcal{S}(T)\}$ denote the class of admissible weak solutions defined on the strip $\mathcal{S}(T)$. Let u and v be two solutions in K whose initial data coincide at almost all x. The problem is to prove that u and v coincide at almost all (x,t) in $\mathcal{S}(T)$. Here we consider a somewhat less general problem. Our results are concerned with the class PL of admissible piecewise Lipschitz solutions. More precisely, PL denotes the class of solutions u in K with the following property. For each t in [0,T) there exists a set of isolated points $\{x_j(t)\}$ such that the restriction $u(\cdot,t)$ of u to each interval (x_j,x_{j+1}) is a Lipschitz function of x; the dependence of the Lipschitz norm on (x_j, x_{j+1}) is arbitrary as well as the dependence of the partition points x_j on t. We note that PL forms a broad subclass of K; PL contains the classical piecewise smooth solutions, i.e. solutions consisting of isolated shock waves, centered and noncentered, rarefaction waves and compression waves and their interactions. In particular PL contains the classical solution of the Riemann problem [20]. We shall discuss the relationship between PL and K in more detail below. We refer the reader to Greenberg [12,13,14] for the construction and analysis of interactions in piecewise smooth solutions and to Oleinik [29], Godunov [11], Rozhdestvenskii [31], Hurd [16,17] and Liu [22,23] for results which establish that certain types of piecewise smooth solutions are equal if their data are equal.

We first consider genuinely nonlinear systems, i.e. systems all of whose eigenvalues are genuinely nonlinear. In the setting of genuinely nonlinear systems of two equations and solutions with small oscillation we establish the following theorem.

Theorem 1. For every state $\tilde{u} \in R^2$ there exists a constant $\delta > 0$ depending only on $\tilde{u}$ and f with the following property. If $w \in PL$, $u \in K$, $|w(\cdot,\cdot)-\tilde{u}|_\infty < \delta$, $|u(\cdot,\cdot)-\tilde{u}|_\infty < \delta$ and $w(x,0) = u(x,0)$ for almost all x then $w(x,t) = u(x,t)$ for almost all (x,t).

We note that the restriction to solutions with small oscillation is not essential for the proof of Theorem 1. By the same method we consider the quasilinear wave equation

$$(12) \qquad \begin{aligned} u^1_t + p(u^2)_x &= 0 \\ u^2_t - u^1_x &= 0 \quad , \end{aligned}$$

with $p' < 0$, $p'' > 0$ and establish uniqueness for solutions with large oscillation.

Theorem 2. If w and u are arbitrary solutions of (12) which lie in PL and K respectively then w and u coincide at almost all (x,t) if their initial data coincide at almost all x.

We recall that under the hypotheses $p' < 0$ and $p'' > 0$, (12) forms a genuinely nonlinear system with a globally defined strictly convex entropy. The class of equations represented by (12) includes the isentropic equations of fluid dynamics, the equations of shallow water waves, and the equations of motion for certain elastic beams. Our interest in (12) also stems from the fact that it serves as the prototype for the broad class of genuinely nonlinear systems introduced by Smoller and Johnson [32]. We expect that Theorem 2 will extend to the general system in the Smoller-Johnson class.

In the case of systems with linearly degenerate eigenvalues we have obtained some preliminary results. We prove for systems of two equations with a strictly convex entropy and eigenvalues satisfying (10) that the classical solution to the Riemann problem is unique within K at least in the case of solutions with small oscillation.

We note that the uniqueness problem for conservation laws is a local problem in space and time. We establish the appropriate local versions of the above results with the aid of the notion of generalized characteristic introduced by Dafermos [4]. Our method is applicable in principle to systems of n equations, but it appears that an additional a priori estimate will be needed to treat the case $n > 2$. The proofs of the above results will be published elsewhere.

Remark 1. In treating the uniqueness problem for arbitrary systems of equations (1) with eigenvalues of the form (10) and solutions with large oscillation it will be necessary to supplement the entropy criterion with additional restrictions in order to rule out all unstable solutions. It is a significant open problem to determine (even formally) what these restrictions should be.

Remark 2. The classes PL and K may be compared in terms of the shock sets of their member solutions. Consider a weak solution to a genuinely nonlinear system and let Ω be a domain in x-t plane which does not contain any shock waves of u. Specifically, we shall refer to a domain Ω as shock-free if

$$H_1(\Omega \cap \Gamma(u)) = 0 \quad .$$

In a forthcoming paper on the regularity of solutions we shall establish the following theorem: let u be a solution in K to a genuinely nonlinear system of two equations; if Ω is an open shock-free domain for u then u is Lipschitz continuous on each compact subdomain of Ω after a possible modification on a set with H_2 measure zero. We conjecture that this result holds for genuinely nonlinear systems of n equations. Thus, in the case of genuinely nonlinear systems (of two equations), PL is essentially the class of admissible weak solutions which have isolated shock waves. In general the shock set may contain points of accumulation. It would be interesting to generalize Theorems 1 and 2 to the case where both u and w contain accumulating shock waves, i.e. to the case where both u and w are arbitrary solutions in K.

We also establish results on the stability of Lipschitz solutions. Let $L = L\{\mathcal{S}(T)\}$ denote the class of Lipschitz solutions defined on the strip $\mathcal{S}(T)$. We prove that solutions in L are L^2-stable relative to perturbations in K.

Theorem 3. Suppose that (1) is a system of n equations with a strictly convex entropy. If $w \varepsilon L\{\mathcal{S}(T)\}$, $u \varepsilon K\{\mathcal{S}(T)\}$ and $0 \leq t < T$ then for all $M > 0$

$$\int_{|x|\leq M} |u(x,t)-w(x,t)|^2 dx \leq c_2 \int_{|x|\leq M+c_1 t} |u(x,0)-w(x,0)|^2 dx \quad .$$

The constant c_1 depends on f, and the L^∞- norms of u and w; the constant c_2 depends on f, T, the L^∞-norms of u and w and the Lipschitz norm of w. We note that in Theorem 3 no hypotheses are placed on (1) beyond the smoothness of f and the existence of a strictly convex entropy. We also establish the corresponding version of Theorem 3 for systems in several space dimensions which have a strictly convex entropy, e.g. fluid dynamics, magneto-fluid dynamics and symmetric hyperbolic systems.

The entropy criterion is illuminated by comparison with other admissibility criteria. In [2,3] Dafermos postulates an entropy rate criterion that identifies the relevant solutions as those which dissipate entropy at the highest possible rate. The entropy rate criterion was introduced by

Dafermos mainly for the purpose of characterizing the class of physically relevant solutions to systems (1) whose eigenvalues are neither genuinely nonlinear nor linearly degenerate.

It is not difficult to prove that the entropy rate criterion is stronger than the entropy criterion: any solution in $L^\infty \cap BV$ which satisfies the entropy rate criterion necessarily satisfies the entropy criterion. It is interesting to consider the converse, i.e. to consider a weak solution which satisfies the entropy criterion and determine in what class (if any) it dissipates entropy at the highest possible. In this direction we prove that for general systems of n equations, all Lipschitz solutions satisfy the entropy rate criterion relative to a broad class of solutions. We also establish a similar result for PL solutions to the quasilinear wave equation (12).

REFERENCES

1. Bakhvarov, N., On the existence of regular solutions in the large for quasilinear hyperbolic systems, Zhur. Vychisl. Mat. i Mathemat. Fiz., 10 (1970), 969-980.
2. Dafermos, C. M., The entropy rate admissibility criterion for solutions of hyperbolic conservation laws, J. Differential Eq., 14 (1973), 202-212.
3. Dafermos, C. M., The entropy rate admissibility criterion in thermoelasticity, Rend. Accad. Naz. dei Lincei, Ser. 8, 64 (1974), 113-119.
4. Dafermos, C. M., Generalized characteristics and the structure of solutions of hyperbolic conservation laws, to appear in Indiana Math. J.
5. DiPerna, R. J., Global solutions to a class of nonlinear hyperbolic systems of equations, Comm. Pure Appl. Math., 26 (1973), 1-28.
6. DiPerna, R. J., Existence in the large for nonlinear hyperbolic conservation laws, Arch. Rational Mech. Anal., 52 (1973), 244-257.
7. DiPerna, R. J., Decay of solutions of hyperbolic systems of conservation laws with a convex extension, Ibid., 64 (1977), 1-46.

8. Federer, H., "Geometric Measure Theory". New York: Springer 1969.
9. Friedrichs, K. O. & P. D. Lax, Systems of conservation laws with a convex extension, Proc. Nat. Acad. Sci. USA, 68 (1971), 1686-1688.
10. Glimm, J., Solutions in the large for nonlinear hyperbolic systems of equations, Comm. Pure Appl. Math., 18 (1965), 697-715.
11. Godunov, S. K., On the uniqueness of the solution of the equations of hydrodynamics, Math. USSR Sb. 40 (1956), 467-478.
12. Greenberg, J. M., On the interaction of shocks and simple waves of the same family II, Arch. Rational Mech. Anal. 37 (1972), 209-277.
13. Greenberg, J. M., Decay theorems for stopping-shock problems, J. Math. Anal. Appl., 50 (1975), 314-324.
14. Greenberg, J. M., Estimates for fully developed shock solutions to the equation, $\partial u/\partial t - \partial v/\partial x = 0$ and $\partial v/\partial t - \partial \sigma(u)/\partial x = 0$, Indiana Univ. Math. J. 22 (1973), 989-1003.
15. Greenberg, J. M., The Cauchy problem for the quasilinear wave equation, unpublished.
16. Hurd, A. E., A uniqueness theorem for weak solutions of symmetric quasilinear hyperbolic systems, Pacific J. Math., 28 (1969), 555-559.
17. Hurd, A. E., A uniqueness theorem for second order quasilinear hyperbolic equations, Ibid., 32 (1970), 415-427.
18. Keyfitz, B., Solutions with shocks, an example of an L^1-contractive semi-group, Comm. Pure Appl. Math., 24 (1971), 125-132.
19. Kruzkov, N., First order quasilinear equations in several independent variables, Math. USSR Sb., 10 (1970), 217-243.
20. Lax, P. D., Hyperbolic systems of conservation laws, II, Comm. Pure Appl. Math. 10 (1957), 537-566.
21. Lax, P. D., Shock waves and entropy, "Contributions to Nonlinear Functional Analysis", ed. E. A. Zarantonello, 603-634. New York: Academic Press 1971.

22. Liu, T.-P., Existence and uniqueness theorems for Riemann Problems, Trans. Amer. Math. Soc. 213 (1975), 375-382.

23. Liu, T.-P., Uniqueness of weak solutions of the Cauchy problem for general 2×2 conservation laws, J. Differential Eq., 20 (1976), 369-386.

24. Liu, T.-P., Solutions in the large for the equations of nonisentropic gas dynamics, Indiana Univ. Math. J., 26 (1977), 147-177.

25. Liu, T.-P., Initial-boundary value problems for gas dynamics, Arch. Rat. Mech. Anal., 64 (1977), 137-168.

26. Nishida, T., Global solutions for an initial boundary value problem of a quasilinear hyperbolic system, Proc. Japan Acad., 44 (1968), 642-646.

27. Nishida, T. & J. A. Smoller, Solutions in the large for some nonlinear hyperbolic conservation laws, Comm. Pure Appl. Math., (1973), 183-200.

28. Oleinik, O. A., Discontinuous solutions of non-linear differential equations, Uspekhi Mat. Nauk (N.S.), 12 (1957), no. 3 (75), 3-73. (A.M.S. Transl., Ser. 2, 26, 95-172.)

29. Oleinik, O. A., On the uniqueness of the generalized solution of the Cauchy problem for a nonlinear system of equations occurring in mechanics, Ibid., 12 (1957), no. 6 (78), 169-176.

30. Oleinik, O. A., Uniqueness and stability of the generalized solution of the Cauchy problem for a quasi-linear equation, Ibid., 14 (1959), 165-170.

31. Rozhdestvenskii, Discontinuous solutions of hyperbolic systems of quasi-linear equations, Uspekhi Mat. Nauk., 15 (1960), no. 6 (96), 59-117 (Russian Math. Surveys, 15 (1960), no. 6, 55-111).

32. Smoller, J. A. & J. L. Johnson, Global solutions for an extended class of hyperbolic systems of conservation laws, Arch. Rat. Mech. Anal., 32 (1969), 169-189.

33. Vol'pert, A. I., The spaces BV and quasilinear equations, Math. USSR Sb. 2 (1967), 257-267.

34. Wendroff, B., The Riemann problem for materials with nonconvex equations of state I: isentropic flow, J. Math. Anal. Appl. 38 (1972), 454-466.

35. Wendroff, B., The Riemann problem for materials with nonconvex equations of state II: general flow, Ibid., 38 (1972), 640-658.

Department of Mathematics
University of Wisconsin-Madison
Madison, Wisconsin 53706

Computational Aspects of Glimm's Method

Alexandre Joel Chorin

1. Introduction and general considerations.

In [9] Glimm introduced an approximate method for constructing solutions of systems of nonlinear hyperbolic conservation laws. This construction is the basis for his beautiful existence theorem in the large (with restrictions on the type of systems allowed and on the size and variation of the data). Some interesting generalizations of Glimm's proof have become available (see e.g. [7], [8], [11, [13]). Since 1965 there have been a number of attempts to use Glimm's construction as an computational tool, but they have not been brought to completion (or to publication), because it is quite clear that in most problems Glimm's construction is less accurate and more expensive than alternative methods. It has, however, turned out in the last few years that there are problems in which Glimm's construction is in fact very useful as a practical tool. These problems involve flow with multiple phases and/or chemical reactions, and the reasons for the usefulness of Glimm's method will become apparent in the course of the discussion. It also turns out that the method is being used under conditions where the assumptions in the available proofs are not satisfied, and a host of interesting open questions are in need of answers.

We begin by describing the method briefly. Consider the hyperbolic system of equations

$$(1) \qquad \underline{v}_t = (\underline{f}(\underline{v}))_x, \quad \underline{v}(x,0) \quad \text{given},$$

where $\underline{v}$ is the solution vector, and subscripts denote dif-

ISBN 0-12-195250-9

ferentiation. The time t is divided into intervals of length k. Let h be a spatial increment. The solution is to be evaluated at the points (ih,nk) and $((i+\frac{1}{2})h, (n+\frac{1}{2})k)$, $i = 0, \pm 1, \pm 2, \ldots$, $n = 1,2,\ldots$. Let $\underline{u}_i^n$ approximate $\underline{v}(ih,nk)$, and $\underline{u}_{i+1/2}^{n+1/2}$ approximate $\underline{v}((i+\frac{1}{2})h,(n+\frac{1}{2})k)$. The algorithm is defined if $\underline{u}_{i+1/2}^{n+1/2}$ can be found when $\underline{u}_i^n$, $\underline{u}_{i+1}^n$ are known. Consider the following Riemann problem:

$$\underline{v}_t = (f(\underline{v}))_x\ , \quad t > 0, \quad -\infty < x < +\infty,$$

$$\underline{v}(x,0) = \begin{cases} \underline{u}_{i+1}^n & \text{for } x \geqslant 0, \\ \underline{u}_i^n & \text{for } x < 0. \end{cases}$$

Let $\underline{x}(x,t)$ denote the solution of this problem. Let θ_i be a value of a variable θ, $-\frac{1}{2} \leqslant \theta \leqslant \frac{1}{2}$. Let P_i be the point $(\theta_i h, \frac{k}{2})$, and let

$$\underline{\tilde{w}} = \underline{w}(P_i) = \underline{w}(\theta_i h, \tfrac{k}{2})$$

be the value of the solution $\underline{w}$ of the Riemann problem at P_i. We set

$$\underline{u}_{i+1/2}^{n+1/2} = \tilde{w}.$$

In other words, at each time step, the solution is first approximated by a piecewise constant function; it is then advanced in time exactly, and new values on the mesh are obtained by sampling.

It is clear that the applicability of the method depends on one's ability to find appropriate allowable solutions of the Riemann problem. The systems under consideration in practice are simplified versions of very complex systems; the omitted equations or terms survive in the form of algebraic constraints on the allowable waves. These constraints define what is allowable. It is one of the major attractions of Glimm's method that it allows one to impose the appropriate constraints with elegant ease.

The rate of convergence of the method (i.e. the error as a function of k and h) depends on how judiciously the θ's are chosen. The θ's must tend to equipartition on $[-1/2, 1/2]$. Various strategies for choosing the θ's are described in [2]; they lead to a method which is almost of first order accuracy. It is clear that the method cannot be more accurate than first order, since, depending on the choice of θ, a value of the solution may become attached to either one of two neighboring points. A more important quantity is the resolution of the method, i.e. its ability to conserve information above waves and their interactions without smoothing. It is shown in [2] that the method has high resolution under conditions where its natural competitors are fitting methods, i.e. methods which treat discontinuities as if they were internal boundaries.

2. Glimm's method in gas dynamics.

I intend to explain in some detail how Glimm's method is used in the analysis of reacting gas flow in one dimension. As a first step, we discuss the flow of chemically inert polytropic gas. The equations of motion are

$$\rho_t + (\rho v)_x = 0 \tag{2a}$$

$$(\rho v)_t + (\rho v^2 + p)_x = 0 \tag{2b}$$

$$e_t + ((e + p)v)_x = 0 \tag{2c}$$

where ρ is the density of the gas, v is its velocity, p is the pressure and e is the energy per unit volume; we have

$$e = \rho\varepsilon + \tfrac{1}{2}\rho v^2 \tag{2d}$$

where ε, the internal energy per unit mass, is given by

$$\varepsilon = \frac{1}{\gamma-1}\frac{p}{\rho} \tag{2e}$$

where $\gamma > 1$ is a constant. Glimm's method rests on the solution of the Riemann problem for these equations. Its convergence has been studied in [12] under the assumption that the data have small variation.

Given a right state $S_r(\rho_r, p_r, v_r, e_r)$ and a left state $S_\ell(\rho_\ell, p_\ell, v_\ell, e_\ell)$, the solution of the Riemann problem consists of, from right to left: The right state S_r, a right wave which either a shock or a rarefaction, a middle right state S_{*r}, a slip line across which p and v (but not necessarily ρ or e) are continuous, a middle left state $S_{*\ell}$, a left wave which is either a shock or a rarefaction, and the left state S_ℓ. Let p_{*r} (resp. $p_{*\ell}$) be the pressure in S_{*r} (resp. $S_{*\ell}$). If $p_{*r} \leqslant p_r$ the right wave is a rarefaction, and if p_{*r} is known, S_{*r} can be determined from S_r, p_{*r}, the constancy of the appropriate Riemann invariant, and the isentropic equation of state. If $p_{*r} > p_r$ the right wave is a shock. The well-known fact that a shock is supersonic with respect to gas in front of it and subsonic with respect to the gas behind it is in fact the condition which ensures that S_{*r} behind a shock can be uniquely determined if S_r and p_{*r} are known. This can be readily seen by drawing and counting the characteristics entering the shock, see e.g. [6]. Similar considerations apply to the determination of $S_{*\ell}$ when $p_{*\ell}$ and S_ℓ are given. The condition $p_{*r} = p_{*\ell}$ ensures that the number of equations equals the number of unknowns. The resulting system of equations has a unique real solution, and can be solved e.g. by Godunov's iteration ([10], see also [14], [1], [2]).

Boundary conditions in a mixed initial value boundary value problem are readily imposed once it is noticed that the slip line separates the fluid initially to its right from the fluid initially to its left. One then designs Riemann problems at the walls in which the slip line has the appropriate velocity (see e.g. [3]). These considerations completely define Glimm's method for a one dimensional inert gas.

3. Glimm's method and the Riemann solution for nonconducting rapidly reacting gas.

We now consider the flow of a gas in which chemical reactions are taking place very rapidly and the heat conduction and viscosity are zero. The equations of motion are

(3a) $$\rho_t + (\rho v)_x = 0$$

(3b) $$(\rho v)_t + (\rho v^2 + p)_x = 0$$

(3c) $$e_t + ((e + p)v)_x = 0$$

(3d) $$e = \rho\varepsilon + \frac{1}{2}\rho v^2$$

as in (2), but now

(3e) $$\varepsilon = \varepsilon_i + q$$

where ε_i is the interval energy per unit mass,

(3f) $$\varepsilon_i = \frac{1}{\gamma_0 - 1}\frac{p}{\rho} \quad \text{for unburned gas,}$$

(3g) $$\varepsilon_i = \frac{1}{\gamma_1 - 1}\frac{p}{\rho} \quad \text{for burned gas,}$$

and q is the energy of formation which can be released through chemical reaction, $q = q_0$ for unburned gas, $q = q_1$ for burned gas. If the chemical reaction is exothermic (i.e. gives out heat) and $\gamma_1 = \gamma_0$, then $\Delta = q_1 - q_0 < 0$. We shall assume that $\gamma_1 = \gamma_0 = \gamma$ to reduce the amount of writing.

The chemical reaction is assumed to occur very fast, and the reaction zone is idealized a s a discontinuity. The jump condition across a discontinuity can be readily derived (see e.g. [6]), and some manipulation leads to the following identities:

(4a) $$-M = \rho_1(v_1 - U) = \rho_0(v_0 - U),$$

(4b) $$M^2 = -\frac{p_0 - p_1}{\tau_0 - \tau_1},$$

and

(4c) $$2\mu^2 H = p_0(\tau_0 - \mu^2\tau_1) - p_1(\tau_1 - \mu^2\tau_0) - 2\mu^2\Delta,$$

where $\tau \equiv 1/\rho$, the index 0 refers to unburned gas, the index 1 refers to burned gas, $\mu^2 = \frac{\gamma-1}{\gamma+1}$, and U is the velocity of the discontinuity. Equation (4c) describes the locus of states (τ_1,p_1) which can be connected to (τ_0,p_0) by an infinitely thin reaction zone with energy release Δ. (See Fig. 1). The curve $H = 0$ is called the Hugoniot curve. The lines through (τ_0,p_0) tangent to $H = 0$ are called the Rayleigh lines and the points of tangency S_1 and S_2 are called the Chapman-Jouguet (CJ) points.

One portion of the Hugoniot curve is omitted because it would yield complex values of M (equation (4b))-clearly an impossibility. The upper branch of the curve is called the detonation branch; the part above S_1 the strong detonation branch, the part below S_1 the weak detonation branch. The lower part of the curve is the deflagration (or flame) branch. The gas burns if the temperature $T = p/\rho$ reaches a value above a critical temperature ("ignition temperature") T_c. This burning is irreversible and occurs at most once for each fluid particle.

Consider a right state $S_r(\rho_r,p_r,v_r,e_r)$ in which the gas is not burned, i.e. $q = q_0$, and consider the states S to which it can be connected by a shock, a rarefaction, or a reaction zone. If $T = p/\rho$ in S is such that no burning occurs, then we are back in the situation of the preceding section, and a knowledge of the pressure p in S will determine S. Suppose T in S is such that burning does occur. If p in S is given and is above the pressure in the Chapman-Jouguet state S_1, one can show that the geometry of the characteristics is such that p in S and S_r will determine S, as would have been the case with a shock. However, if p in S is below p in S_1 the situation is more complex. Weak detonations more more rapidly then the sound speed behind them (see [6]) and the state behind them is undetermined. Fortunately, one can show that weak detonations do not occur if the energy release is due to a chemical reaction in which the proportion of fuel burned remains between 0 and 100%. What can occur is a CJ detonation. The velocity of CJ detonation S_1 equals the speed of sound behind it, and it can thus be

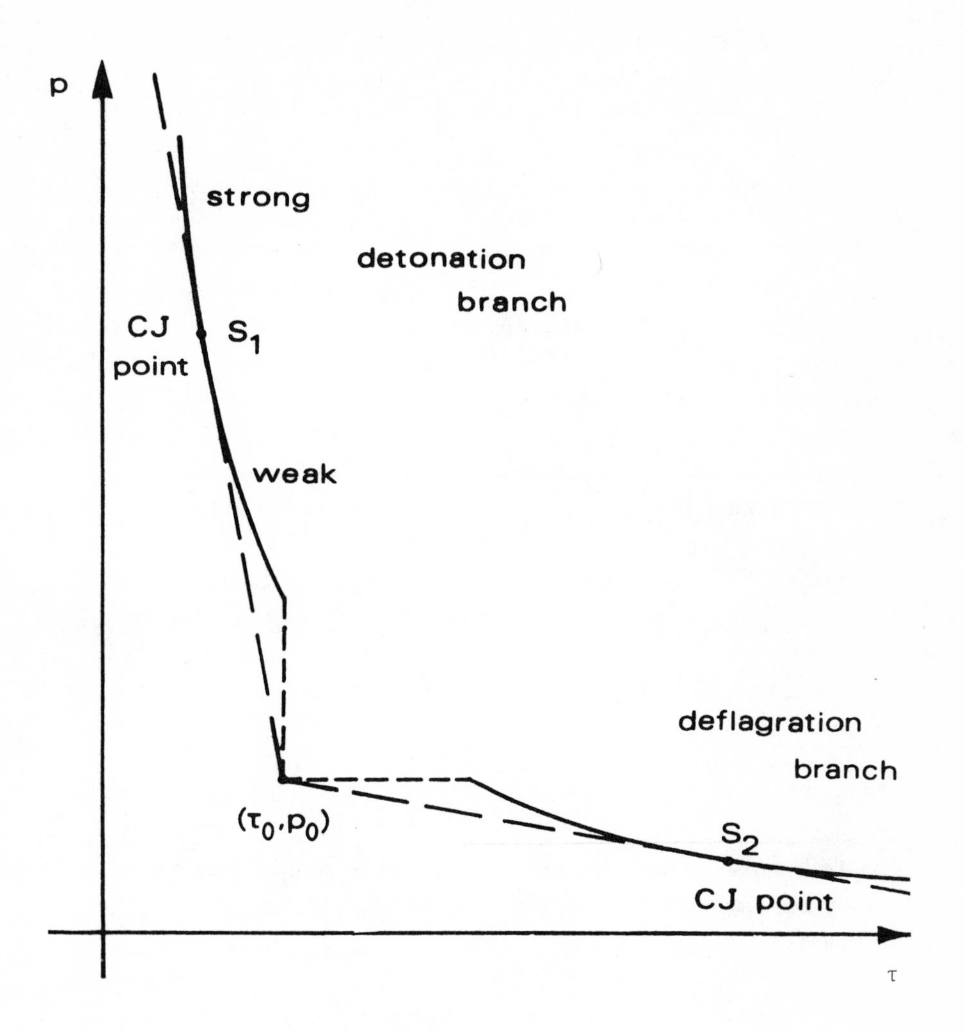

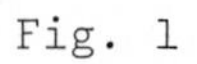
Fig. 1

The Hugoniot Curve

followed by a rarefaction. S_1 is entirely determined by S_r, and if p in S is given, it determines S by connecting it to S_1 by a rarefaction ([6], [2]).

One can further show that in the absence of heat conduction the only deflagration which can occur is one across which p is constant and there is no mass flow -- i.e., such a deflagration is indistinguishable from a slip line (see e.g. [2]). Thus, the Riemann problem for equations (3) can be reduced to the Riemann problem for an inert gas, as long as one allows the right and left waves to be not only rarefactions or shocks, but also strong detonations or CJ detonations followed by rarefactions. A shock is the special case of a strong detonation with $\Delta = 0$, and a CJ detonation followed by a rarefaction is merely a rarefaction connected to a state S_1 entirely determined by either S_r on the right or S_ℓ on the left. Godunov's algebraic iteration procedure for solving the Riemann problem easily generalizes to the reacting case ([2]).

It is interesting to note that the elimination of parts of the Hugoniot curve by physical constraints, which makes the Riemann problem determinate, has a strong qualitative resemblance to the procedures used to define an appropriate Riemann solution for problems of the form $u_t + f_x$, where f is a non-convex function of u (see e.g. [7]), or to gas flow problems where the gas has a nonconvex equation of state ([16]).

Furthermore, the discussion just given should put in evidence some of the practical advantages of Glimm's method. Consider a problem whose solution consists of a CJ detonation followed by a rarefaction moving down a tube. The velocity and pressure profiles in such a wave are not monotonic, and may contain a very sharp spike at the detonation. Such a wave is very difficult to describe by a finite difference method, but will be described nearly perfectly by Glimm's method.

4. Reacting gas flow with heat conduction and finite rate chemistry.

We now consider how the method just described can be generalized to the case of flow with realistic chemistry and finite heat conduction. Finite heat conduction is important since one of the main mechanisms of flame (deflagration) propagation is the conduction of heat which ignites new fluid (the other major mechanism is the diffusion of chemical species which can be treated by similar methods but which we shall not consider here). The full equations which describe the flow are enormously complex (see e.g. [17]) and must be simplified before they can be used or else the calculation becomes prohibitively expensive and the results impossible to interpret. An appropriate simplified model is:

(5a) $$\rho_t + (\rho v)_x = 0$$

(5b) $$(\rho v)_t + (\rho v^2 + p)_x = 0$$

((5a,b) are identical to (3a,b)),

(5c) $$e_t + ((e + p)v)_x - \lambda T_{xx}, \quad T = p/\rho ,$$

(5d) $$e = \rho\varepsilon + \tfrac{1}{2}\rho v^2,$$

(5e) $$\varepsilon = \frac{1}{\gamma-1}\frac{p}{\rho} + Zq, \quad 0 \leqslant Z \leqslant 1,$$

(5f) $$Z = Z(\underline{s}), \ \gamma = \gamma(\underline{s}), \ \underline{s} = (s_1,\ldots,s_n),$$

(5g) $$\frac{d\underline{s}}{dt} = G(T,\underline{s})$$

where λ is the heat conduction, q is the total available chemical energy, Z is the fraction of energy still not released, and equation (4f) and (5g) express the fact that Z is a function of the progress of a chemical reaction which is described by a suitable kinetic scheme. A simple model of (5f), (5g) is the set of equations

$$\frac{dZ}{dt} = -KZ, \qquad \gamma = \text{constant},$$

where $K = 0$ if $T = p/\rho \leqslant T_0$, $K = K_0 = \text{constant} > 0$ if $T > T_0$. K_0 is the reaction rate and T_0 is the ignition temperature. If λ is not very small and K_0 is not very large, the added equations can be handled by a standard fractional step procedure and taken outside the Riemann problem. However, if λ is very small and K_0 is large one has to find a suitable approximate Riemann solution which takes their effects into account.

The remarkable fact is that this can be done; it can be done most economically by making the following approximations: in the case of a detonation of finite structure, the state behind the detonation is the CJ state. In a case of a deflagration, there is no change of pressure across the deflagration. These approximations are in fact in very good agreement with experimental fact and with the relevant theory, and allow one to use the construction of the preceding section almost without change. For details, see e.g. [4], [15]. As a result, Glimm's method is capable of handling flows which contain hydrodynamic effects and chemical effects of widely different time scales.

For another type of application of Glimm's method, see [5].

REFERENCES

[1] A. J. Chorin, J. Comp. Phys., 22, 517 (1976).

[2] A. J. Chorin, Random choice methods with applications to reacting gas flow, to appear in J. Comp. Phys.

[3] A. J. Chorin, Vortex sheet approximation of boundary layers, to appear in J. Comp. Phys.

[4] A. J. Chorin, A numerical model of flame propagation, to appear.

[5] P. Concus and W. Proskurowski, Numerical solution of a nonlinear hyperbolic equation by the random choice method, to appear.

[6] R. Courant and K. O. Friedrichs, Supersonic flow and shock waves, Interscience (1948).

[7] C. Dafermos, Arch. Rat. Mech. Anal., 52, 1, (1973).

[8] R. Diperna, Comm. Pure Appl. Math., 26, 1 (1973).

[9] J. Glimm, Comm. Pure Appl. Math. 18, 697 (1965).

[10] S. K. Godunov, Mat. Sbornik, 47, 271 (1959).

[11] N. N. Kuznetsov and V. A. Tupshiev, Dokl. Acad. Nauk USSR, 221, 287 (1975).

[12] T. P. Liu, Indiana Univ. Math. J., 26, 147 (1977).

[13] T. P. Liu, A deterministic version of Glimm's method, to appear.

[14] R. D. Richtmyer and K. W. Morton, Finite Difference Methods for Initial Value Problems, Interscience (1967).

[15] G. Sod, to appear.

[16] B. Wendroff, J. of Math. 38, 454 (1972).

[17] F. A. Williams, Combustion theory, Addison Wesley (1965).

Partially supported by US ERDA at the Lawrence Berkeley Laboratory.

Department of Mathematics
University of California
Berkeley, California 94720

The Initial Value Problem of the Boltzmann Equation and Its Fluid Dynamical Limit at the Level of Compressible Euler Equation

Takaaki Nishida

1. INTRODUCTION

The Boltzmann equation for the mass density distribution function $F = F(t,x,v)$ describes the motion of rarefied gases

$$(1.1) \qquad \frac{\partial F}{\partial t} + v\cdot \frac{\partial F}{\partial x} = \frac{1}{\varepsilon} Q(F,F),$$

where $t \geq 0$: time, $x \in R^3$: physical space, $v \in R^3$: velocity space, ε is the mean free path and Q represents the binary collision of the gas molecules :

$$(1.2) \quad Q(F,G) = \frac{1}{2}\int (F'G'_* + F'_*G - FG_* - F_*G)\, Vr\, dr\, d\phi\, dv_* ,$$

where $V = |v-v_*|$, v' and v'_* are the velocities after the collision of the molecules with the velocities v and v_*, r and ϕ are the polar coordinate in the impact plane, $F_* = F(t, x,v_*)$, $F' = F(t,x,v')$, $F'_* = F(t,x,v'_*)$ and the same notation is used for G. The initial data are given by

$$(1.3) \qquad F(0,x,v) = F(x,v) \geq 0.$$

We assume the cut-off hard potentials in the sense of Grad (1965) for the collision, which include the hard sphere

ISBN 0-12-195250-9

molecules. The initial value problem (1.1) (1.3) was investigated locally in time by Grad (1965) and globally in time by Ukai (1974, 1976), Nishida and Imai (1976) and then Shizuta (to appear). If the initial deviation from the absolute Maxwellian distribution

$$(1.4) \qquad M(v) = \frac{1}{(2\pi)^{3/2}} e^{-v^2/2}$$

is small with ε fixed, then the solution is proved to exist uniquely in the large in time and to converge to the absolute Maxwellian as time goes to infinity. The spectral theory for the linear Boltzmann equation (see Ellis and Pinsky (1975)) gives the decay estimates on solutions of the linear Boltzmann equation ; and an arguement of Grad (1965) gives solutions to the full Boltzmann equation as a nonlinear perturbation from the linear equation.

The initial-boundary value problems (1.1) in the bounded domains are solved globally in time by Guiraud (1974) for the boundary condition of random reflection and by Shizuta and Asano (1977) for the specular boundary condition.

Let us define the summational invariants.

$$(1.5) \qquad \{\Psi_j\}_{j=1}^{5} = \{1, v_j \ (j=1,2,3), v^2\}$$

which satisfy

$$(1.6) \qquad \int \Psi_j \, Q(F,G)\, dv = 0 \ , \quad j=1,2,\cdots, 5$$

by virtue of the conservation of mass, momentum and energy under the collision. The hydrodynamical quantities are defined as follows :

$$(1.7) \qquad \rho(t,x) \equiv \int F(t,x,v)\, dv : \text{mass density,}$$

(1.8) $u(t,x) \equiv \frac{1}{\rho}\int vF(t,x,v)dv$: fluid flow velocity.

Using the velocity relative to the mean $c = v-u(t,x)$ the stress tensor and heat-flow vector are defined by

$$(1.9)\qquad P_{ij} \equiv \int c_i c_j F(t,x,v)dv = p_{ij} + p\delta_{ij} ,$$

$$(1.10)\qquad q_i \equiv \frac{1}{2}\int c_i c^2 F(t,x,v)dv ,$$

where $p = \frac{1}{3}P_{jj}$ is the scalar pressure. The internal energy per unit mass is

$$(1.11)\qquad e \equiv \frac{1}{\rho}\int \frac{1}{2}c^2 F(t,x,v)dv .$$

Multiplication (1.1) by Ψ_j, $j=1,2,\cdots,5$ and integration in v give the conservation laws for ρ, u, e :

$$\rho_t + (\rho u_j)_{x_j} = 0$$

$$(1.12)\qquad (\rho u_i)_t + (\rho u_i u_j + p\delta_{ij} + p_{ij})_{x_j} = 0$$

$$(\rho(e+u^2/2))_t + (\rho u_j(e+u^2/2)+pu_j+p_{kj}u_k+q_j)_{x_j} = 0$$

where the equation of state of gas is that of the ideal gas i.e.,

$$(1.13)\qquad RT \equiv p/\rho = \frac{2}{3}e .$$

The system (1.12) is not closed by itself, because for p_{ij} and q_i we need the higher momentum of F. But if the distribution function F is locally Maxwellian i.e.,

$$(1.14)\qquad F(t,x,v) = \frac{\rho(t,x)}{(2\pi RT(t,x))^{3/2}}\exp\left(-\frac{(u(t,x)-v)^2}{2RT(t,x)}\right),$$

then the conservation laws (1.12) can be simplified by $p_{ij} = q_i = 0$ to

$$(1.15)\quad \begin{aligned} &\rho_t + (\rho u_j)_{x_j} = 0 \\ &(\rho u_i)_t + (\rho u_i u_j + p\delta_{ij})_{x_j} = 0 \\ &(\rho(e+u^2/2))_t + (\rho u_j(e+u^2/2) + pu_j)_{x_j} = 0\ , \end{aligned}$$

which is now closed with (1.13) and may be considered as the compressible Euler equation derived from the Boltzmann equation and is the same system of hyperbolic conservation laws for the ideal compressible gas motion. It is the first approximation of the Chapman-Enskog procedure or Hilbert expansion. The second approximation of the Chapman-Enskog expansion is the compressible Navier-Stokes equation. The asymptotic problem of the Boltzmann equation (1.1) as the mean free path ε tends to zero and the relations to the hydrodynamical equations by Chapman-Enskog expansion are considered by Grad (1965) for the "semilinear" Boltzmann equation locally in time and by McLennan (1965), Ellis and Pinsky (1975) and Pinsky (1976) for the linear Boltzmann equation. In § 4 we consider the hydrodynamical limit of the nonlinear Boltzmann equation as the mean free path ε goes to zero at the level of compressible Euler equations. If the initial deviation from the absolute Maxwellian is small and analytic in the space variables, the solution of the Boltzmann equation exists in a finite time interval independent of ε and it converges there, as $\varepsilon \to 0$, to the local Maxwellian distribution whose fluid-dynamical quantities satisfy the compressible Euler equation (1.15). (see Nishida (to appear)) This is proved by use of the linear spectral theory and the abstract Cauchy-Kowalewski theorem (see

Nirenberg (1972) and Nishida (to appear)). Last we note that the one-dimensional shock wave solutions of Boltzmann equation are investigated by Nicolaenko and Thurber (1975) and Nicolaenko (1974, preprint).

2. NOTATIONS AND LINEARIZED BOLTZMANN EQUATION

$x, v \in R^3$ are the space - and velocity - variables and $k \in R^3$ is the variable for the Fourier transform in x. $L^p(\cdot)$ ($\cdot = x, v$ or k) denotes the Lebesgue space of measurable functions whose p-th power ($1 \le p < +\infty$) is summable in R^3 with the norm $|f|_{L^p(\cdot)}$. $H^\ell(x)$, $\ell \ge 0$ denotes the Sobolev space of $L^2(x)$-functions together with the ℓ-th derivatives, $\hat{H}^\ell(k)$ is the Fourier transform of $H^\ell(x)$ with the norm

$$|f|_{H^\ell(x)} = |(1+|k|)^\ell \hat{f}(k)|_{L^2(k)} = |\hat{f}|_{\hat{H}^\ell(k)} .$$

Let H be the Lebesgue space of square summable functions in $(x,v) \in R^6$ with the norm

$$\|f\| = \left(\int |f(x,v)|^2 \, dx \, dv \right)^{1/2} . \tag{2.1}$$

Let us introduce the (partial) Fourier transform in x of $f \in H$ by

$$\hat{f}(k,v) = \frac{1}{(2\pi)^{3/2}} \int e^{-ik\cdot x} f(x,v) dx \tag{2.2}$$

and denote $\hat{H} = \{\hat{f} \; ; \; f \in H\}$ with the norm

$$\|\hat{f}\| \equiv \left(\int |\hat{f}(k,v)|^2 \, dk \, dv \right)^{1/2} = \|f\| . \tag{2.3}$$

<u>Definition 2.1.</u> Let the Hilbert space H_ℓ, $\ell \ge 0$ be a subspace of H, which consists of $H^\ell(x)$-valued L^2-functions in $v \in R^3$, i.e.,

$$(2.4)\quad \begin{aligned} H_\ell &= L^2(v;H^\ell(x)) \quad \text{with the norm} \\ \|f\|_\ell &\equiv \left(\int |f(\cdot,v)|^2_{H^\ell(x)} dv\right)^{1/2} \\ &= \left(\int (1+|k|)^{2\ell} |\hat{f}(k,v)|^2 dk\, dv\right)^{1/2} = \|\hat{f}\|_\ell < +\infty . \end{aligned}$$

Also we use the space $L^2(v;L^p(x))$, $1 \le p \le 2$ which consists of $L^p(x)$-valued L^2-function in $v \in R^3$ with the norm

$$(2.5)\quad \|f\|_{L^{2,p}} = \left(\int f(\cdot,v)|^2_{L^p(x)} dv\right)^{1/2} < +\infty ,$$

where $H = L^2(v;L^2(x)) = H_0$.

Definition 2.2. Let $B_{m,\ell}$, $m,\ell \ge 0$ be a subspace of H_ℓ, which consists of $H^\ell(x)$-valued continuous function in v, with the property

$$(2.6)\quad (1+|v|)^m |f(\cdot,v)|_{H^\ell(x)} \to 0 \quad \text{as} \quad |v| \to +\infty .$$

The norm for $f \in B_{m,\ell}$ is defined by

$$(2.7)\quad \begin{aligned} \|f\|_{m,\ell} &= \sup_v (1+|v|)^m |f(\cdot,v)|_{H^\ell(x)} \\ &= \sup_v (1+|v|)^m |\hat{f}(\cdot,v)|_{\hat{H}^\ell(k)} < +\infty . \end{aligned}$$

It is easy to see that by Fubini's theorem

$$(2.8)\quad \|f\|_\ell \le C\|f\|_{m,\ell} \quad \text{for} \quad m > 3/2,\ \ell \ge 0,$$

and that by Sobolev's lemma $f \in B_{m,\ell}$ for $m,\ell > 3/2$ is continuous in x and v.

Definition 2.3. $S_\ell = \bigcup_{\rho \ge 0} H_{\ell,\rho}$ for some $\ell \ge 0$ is a scale of the Hilbert spaces such that $H_{\ell,0} = H_\ell$ and

$$(2.9)\quad H_{\ell,\rho} = \{f \in H_\ell;\ |||f|||_{\ell,\rho} \equiv \|e^{|k|\rho}\hat{f}(k,v)\|_\ell < +\infty\}$$

$S_{m,\ell} = \bigcup_{\rho \ge 0} B_{m,\ell,\rho}$ for some $m,\ell \ge 0$ is a scale of Banach spaces such that $B_{m,\ell,0} = B_{m,\ell}$ and

$$(2.10)\quad B_{m,\ell,\rho} = \{f \in B_{m,\ell};\ |||f|||_{m,\ell,\rho} \equiv \| e^{|k|\rho}\hat{f}(k,v)\|_{m,\ell} < +\infty$$

with the property

$$\lim_{|v|\to+\infty} (1+|v|)^m |e^{|k|\rho}\hat{f}(k,v)|_{\hat{H}^{\ell}(k)} = 0\}$$

Lemma 2.1. The scale of the Hilbert space S_ℓ for any $\ell \geq 0$ has the property

$$(2.11)\quad |||\, |k|^{\sigma}\hat{f}(k,v)\, |||_{\ell,\rho'} \leq \frac{c}{(\rho-\rho')^{\sigma}}\ |||f|||_{\ell,\rho}$$

for any $f \in H_{\ell,\rho}$ and any $\rho'<\rho$, $0<\sigma\leq 1$.

In order to linearize the Boltzmann equation (1.1) around the absolute Maxwellian $M(v)$ we set

$$(2.12)\quad F(t,x,v) = M + M^{1/2} f(t,x,v).$$

If we substitute (2.12) into (1.1) and follow Grad (1963) (1965) for the gas molecules with the cut-off hard potential, we have the equation for $f(t,x,v)$

$$(2.13)\quad \frac{\partial f}{\partial t} + \Sigma\, v_j \frac{\partial f}{\partial x_j} = \frac{1}{\varepsilon}\,(Lf + \nu\Gamma(f,f)).$$

Here L is a nonpositive linear operator acting on $v \in R^3$,

$$(2.14)\quad (Lf,f)_{L^2(v)} \leq 0 \quad \text{for} \quad f, Lf \in L^2(v)$$

and

$$(2.15)\quad Lf = 0 \quad \text{iff} \quad f = \{\Psi_j\}_{j=1}^{5} = \{M^{1/2},\ v_j M^{1/2},\ v^2 M^{1/2}\}.$$

It can be decomposed as

$$(2.16)\quad L = -\,\nu(v) + K \quad \text{in} \quad L^2(v)$$

where $\nu(v)$ is a monotone nondecreasing function in $|v|$ and

$$(2.17)\quad 0 < \nu_0 \leq \nu(v) \leq \nu_1(1+|v|),$$

K is a compact self-adjoint operator on $L^2(v)$, which has the smoothing properties :

$$(2.18)\quad \begin{aligned} &|||Kf|||_{j,\ell,\rho} \leq K\ |||f|||_{j-1,\ell,\rho} \quad \text{for any}\quad j \geq 1, \\ &|||Kf|||_{0,\ell,\rho} \leq K\ |||f|||_{\ell,\rho} \end{aligned}$$

for some constant $K = K(j) < +\infty$ and any $\ell \geq 0$, $\rho \geq 0$.

The nonlinear operator

$$(2.19)\quad \nu\Gamma(f,g) = \frac{1}{2}\int (f'g'_*+f'_*g'-fg_*-f_*g)M(v_*)^{1/2}Vrdrd\phi dv_*$$

acts on $v \in R^3$ and is bilinear in f and g.

Lemma 2.2. Let $f(x,v)$, $g(x,v) \in B_{m,\ell,\rho}$ for some $m>5/2$, $\ell>3/2$ and $\rho \geq 0$. Then we have

$$(2.20)\quad \begin{aligned} |||\nu\Gamma(f,g)|||_{\ell,\rho} &\leq C\ |||\Gamma(f,g)|||_{m,\ell,\rho} \\ &\leq C\ |||f|||_{m,\ell,\rho}\ |||g|||_{m,\ell,\rho} \quad \text{and then} \end{aligned}$$

$$(\nu\Gamma(f,g),\Psi_j)_{L^2(v)} = 0, \quad j = 1,\cdots,5.$$

Proof The first inequality is easily obtained by (2.8) and (2.17). The second is proved by Grad (1965) and by Handsdorff-Young's inequality. In fact for $\ell=2$ we have

$$\begin{aligned} |||\Gamma(f,g)|||_{m,2,\rho} &= \sup_v (1+|v|)^m |(1+|k|)^2 e^{|k|\rho}\hat{\Gamma}(f,g)(k,v)|_{L^2(k)} \\ &\leq C\{\sup(1+|v|)^m\ |e^{|k|\rho}\hat{f}(k,v)|_{L^1(k)}\}\cdot \\ &\quad \cdot \{\sup(1+|v|)^m |(1+|k|)^2 e^{|k|\rho}\hat{g}(k,v)|_{L^2(k)}\} + \\ &\quad + C\{\sup(1+|v|)^m |(1+|k|)^2 e^{|k|\rho}\hat{f}(k,v)|_{L^2(k)}\}\cdot \end{aligned}$$

$$\cdot \{\sup(1+|v|)^m \| e^{|k|\rho}\hat{g}(k,v)\|_{L^1(k)}\}$$

$$\le C\{\sup(1+|v|)^m \| (1+|k|)^2 e^{|k|\rho}\hat{f}\|_{L^2(k)}\} \cdot$$

$$\cdot \{\sup(1+|v|)^m \| (1+|k|)^2 e^{|k|\rho}\hat{g}\|_{L^2(k)}\}$$

$$= C \, |||f|||_{m,2,\rho} \, |||g|||_{m,2,\rho}$$

Now our aim in this section is to summarize some results on the linear Boltzmann equation.

$$\frac{\partial f}{\partial t} = - \Sigma v_j \frac{\partial f}{\partial x_j} + \frac{1}{\varepsilon} Lf. \tag{2.21}$$

Consider two operators

$$\begin{aligned} \frac{1}{\varepsilon} A_\varepsilon &= - \Sigma v_j \frac{\partial f}{\partial x_j} - \frac{1}{\varepsilon} \nu(v), \\ \frac{1}{\varepsilon} \mathbf{B}_\varepsilon &= - \Sigma v_j \frac{\partial f}{\partial x_j} - \frac{1}{\varepsilon} Lf \end{aligned} \tag{2.22}$$

with the domain $D(\frac{1}{\varepsilon} A_\varepsilon) = D(\frac{1}{\varepsilon} B_\varepsilon)$ maximal in H_ℓ, $\ell \ge 0$. $\frac{1}{\varepsilon} A_\varepsilon$ generates a strongly continuous semigroup in H_ℓ, i.e.,

$$\begin{aligned} e^{\frac{t}{\varepsilon} A_\varepsilon} f &= e^{-\frac{t}{\varepsilon}\nu(v)} f(x - \frac{t}{\varepsilon} v, v) \\ &= \frac{1}{(2\pi)^{3/2}} \int e^{ik\cdot x} e^{\frac{t}{\varepsilon} A_{\varepsilon k}} \hat{f}(k,v)\, dk , \end{aligned} \tag{2.23}$$

where

$$A_{\varepsilon k} = - i\varepsilon k\cdot v - \nu(v). \tag{2.24}$$

Since $B_\varepsilon = A_\varepsilon + K$ and K is a bounded perturbation, the linear Boltzmann operator $\frac{1}{\varepsilon} B_\varepsilon$ generates also a strongly continuous semigroup $\{e^{\frac{t}{\varepsilon} B_\varepsilon}\} t \ge 0$ in H_ℓ for any $\varepsilon \in (0,1]$.

Then we have

<u>Theorem 2.1.</u> <u>The linear Boltzmann semigroup is represented by</u>

$$(2.25)\quad e^{\frac{t}{\varepsilon}B_\varepsilon} f = \frac{1}{(2\pi)^{3/2}} \int e^{ik\cdot x} e^{\frac{t}{\varepsilon}B_{\varepsilon k}} \hat{f}(k,v)\, dk$$

$$\text{for}\quad \hat{f}(k,v) \in \hat{H}_\ell ,$$

where for each $k \in R^3$

$$(2.26)\quad B_{\varepsilon k} = - i\varepsilon k.v - \nu(v) + K$$

is a unbounded linear operator in $L^2(v)$ with the definition domain $D(B_{\varepsilon k}) = \{f \in L^2(v),\ B_{\varepsilon k} f \in L^2(v)\}$ and generates a strongly continuous semigroup such that for $f \in L^2(v)$

$$(2.27)\quad |e^{\frac{t}{\varepsilon}B_{\varepsilon k}} f|_{L^2(v)} \le |f|_{L^2(v)} .$$

Furthermore there exist δ, β_1, $\beta_2 > 0$ such taht the following (i) (ii) are valid for any $f \in D(B_{\varepsilon k})$.

(i) for any k, $|\varepsilon k| < \delta$

$$(2.28)\quad e^{\frac{t}{\varepsilon}B_{\varepsilon k}} f = \sum_{j=1}^{5} e^{\frac{t}{\varepsilon}\alpha_j(\varepsilon k)} (e_j(-\varepsilon k), f)_{L^2(v)}\, e_j(\varepsilon k)$$

$$+ e^{\frac{t}{\varepsilon}A_{\varepsilon k}} f + e^{-\frac{t}{\varepsilon}\beta_1} Z_1(\varepsilon k, t/\varepsilon) f ,$$

where α_j, e_j are the eigenvalues and the eigenfunctions of $B_{\varepsilon k}$ such that

$$(2.29)\quad \begin{aligned} \alpha_j(\varepsilon k) &= \sum_{n=1}^{3} a_{j,n}(i\varepsilon|k|)^n + D(|\varepsilon k|^4) \\ e_j(\varepsilon k) &= \sum_{n=0}^{3} e_{j,n}(k/|k|)(i\varepsilon|k|)^n + D(|\varepsilon k|^4), \end{aligned}$$

$a_{j,n}$ are constants, $a_{j,2} > 0$ and

$$(e_j(-\varepsilon k),\ e_n(\varepsilon k))_{L^2(v)} = \delta_{j,n}\ ,\quad j,n = 1,\cdots,5.$$

(ii) for any $k, |\varepsilon k| > \delta$

$$(2.30)\quad e^{\frac{t}{\varepsilon}B_{\varepsilon k}} f = e^{\frac{t}{\varepsilon}A_{\varepsilon k}} f + e^{-\frac{t}{\varepsilon}\beta_2} Z_2(\varepsilon k, t/\varepsilon) f,$$

where

$$
(2.31)\quad \begin{aligned} Z_j(\varepsilon k, t/\varepsilon)f &= \lim_{\gamma\to\infty} \frac{1}{2\pi}\int_{-i\gamma}^{i\gamma} e^{i\frac{t}{\varepsilon}\gamma} Z(-\beta_j + i\gamma,\ \varepsilon k) f\, d\gamma\ , \\ Z(\lambda, \varepsilon k) &= (\lambda - A_{\varepsilon k})^{-1}(I-K(\lambda-A_{\varepsilon k})^{-1})^{-1}\ K(\lambda - A_{\varepsilon k})^{-1} \end{aligned}
$$

and

$$
\| Z_j(\varepsilon k\ , t/\varepsilon) f \|_{L^2(v)} \le C \| f \|_{L^2(v)}\ ,
$$

where C is independent of ε, k, $t \ge 0$.

Proof cf. Ellis-Pinsky (1975), Ukai (1976) and Nishida-Imai (1976).

3. THE INITIAL VALUE PROBLEM OF THE BOLTZMANN EQUATION

First we obtain the decay of solutions to the initial value problem of the linear Boltzmann equation

$$
(3.1)\quad \frac{\partial f(t)}{\partial t} = -\ \Sigma v_j \frac{\partial f}{\partial x_j} + \frac{1}{\varepsilon} Lf \equiv \frac{1}{\varepsilon} B_\varepsilon f\ .
$$

Let the initial data

$$
(3.2)\quad f(0) = f(x,v) \in H_\ell \quad \text{for some} \quad \ell \ge 0.
$$

This Cauchy problem is solved by the linear Boltzmann semi-group in § 2, i.e.,

$$
(3.3)\quad f(t) = e^{\frac{t}{\varepsilon} B_\varepsilon} f, \quad \text{in} \quad t \ge 0,
$$

which is strongly continuous of $t \ge 0$ in H_ℓ. By theorem 2.1 and by Planchrel theorem we have

$$
(3.4)\quad \| f(t) \|_\ell \le \| f \|_\ell \quad \text{in} \quad t \ge 0.
$$

Theorem 3.1.

(i) Let the initials f belong to H_ℓ for some $\ell \ge 0$. Then

<u>the solution</u> $f(t)$ <u>of</u> (3.3) <u>with ε fixed decays to zero</u> :

(3.5) $\quad \| f(t) \|_{\ell} \to 0 \quad$ <u>as</u> $\quad t \to +\infty$

(ii) <u>Let</u> $f \in H_{\ell} \cap L^2(v;L^1(x))$ <u>for some</u> $\ell \geq 0$ <u>and</u>

(3.6) $\quad \int \Psi_j(v) f(x,v)\, dv = 0 \quad$ <u>for a.a.</u>$x \in R^3, \quad j=1,\cdots,5.$

<u>Then the decay estimate has the order as follows</u> :

(3.7) $$\| f(t) \|_{\ell} \leq \frac{C\,(\| f \|_{\ell} + \| f \|_{L^{2,1}})}{(1 + \varepsilon t)^{5/4}} \quad .$$

<u>Remark</u> If $f \in H_{\ell} \cap L^2(v;L^p(x))$ for some $\ell \geq 0$, $2 \geq p \geq 1$, then the decay estimate is better than (i), i.e.,

$$\| \partial_x^{\alpha} f(t) \| \leq \frac{C(\| f \|_{|\alpha|} + \| f \|_{L^{2,p}})}{(1 + \varepsilon t)^{\beta+|\alpha|/2}} \quad \text{for} \quad |\alpha| \leq \ell \ ,$$

where $\partial_x^{\alpha} = (\partial/\partial x_1)^{\alpha_1} \cdots\cdot (\partial/\partial x_n)^{\alpha_n}$, $|\alpha| = \alpha_1 + \cdots + \alpha_n$ and $\beta = \frac{3}{2}\left(\frac{1}{p} - \frac{1}{2}\right)$. And also these estimates can be carried to the solution of nonlinear Boltzmann equation, but here we restrict ourselves to the basic general case (i), (3.5). See Ukai (1976) and Nishida-Imai (1976).

<u>Proof</u> By Fubini theorem and by Planchrel theorem we can compute for $f(t)$

$$\| f(t) \|_{\ell}^2 = \iint (1+|k|)^{2\ell} |e^{\frac{t}{\varepsilon} B_{\varepsilon k}} \hat{f}(k,v)|^2 \, dv\, dk$$

$$= \int_{|\varepsilon k|<\delta} \left(\int (\quad)^2 dv\right) dk + \int_{|\varepsilon k|>\delta} \left(\int (\quad)^2 dv\right) dk \equiv I + I_1,$$

where δ is defined in Theorem 2.1. By Theorem 2.1 (ii) we get the estimate with $\beta_0 = \min(\beta_2, \nu(0))$

$$I_1 \leq C^2 e^{-2\beta_0 t/\varepsilon} \int_{|\varepsilon k|>\delta} (1+|k|)^{2\ell} |\hat{f}(k,\cdot)|^2_{L^2(v)} \, dk$$

$$\leq C^2 e^{-2\beta_0 t/\varepsilon} \| f \|_{\ell}^2 \ ,$$

which means the exponential decay.

By Theorem 2.1 (i) for I we have

$$I = I_2 + I_3 ,$$

where the integrand of I_2 is the first term in the right hand side of (2.28) and that of I_3 is the second and third ones in that of (2.28) respectively. Then Theorem 2.1 (2.28), (2.31) with $\beta_0 = \min(\beta_1, \nu(0))$ gives

$$\begin{aligned} I_3 &\le C^2 e^{-2\beta_0 t/\varepsilon} \int_{|\varepsilon k|<\delta} (1+|k|)^{2\ell} |\hat{f}(k,\cdot)|^2_{L^2(v)} dk \\ &\le C^2 e^{-2\beta_0 t/\varepsilon} \| f \|^2_\ell . \end{aligned}$$

If we set $\alpha_0 = \min\limits_{j=1,2,\cdots,5} \alpha_{j,2} > 0$ in (2.29) we can calculate for I_2 as follows :

$$\begin{aligned} I_2 &= \int_{|\varepsilon k|<\delta} (1+|k|)^{2\ell} |\Sigma e^{\frac{t}{\varepsilon}\alpha_j(\varepsilon k)} (e_j(-\varepsilon k), \hat{f})_{L^2(v)} e_j(\varepsilon k)|_{L^2(v)} dk \\ &\le C^2 \int_{|\varepsilon k|<\delta} (1+|k|)^{2\ell} e^{-t\varepsilon\alpha_0 k^2} |\hat{f}(k,\cdot)|^2_{L^2(v)} dk \\ &\to 0 \quad \text{as} \quad t \to +\infty, \end{aligned}$$

where the decay to zero is assured by Lebesgue theorem. The proof of (ii) is given in the same way, if we note that for $j=1,2,\cdots,5$

$$(e_j(0), \hat{f}(k,\cdot))_{L^2(v)} = \frac{1}{(2\pi)^{3/2}} \int e^{-ik\cdot x} \left(\int \Psi_j f(x,v)\, dv \right) dx = 0$$

and that

$$\begin{aligned} I_2 &\le C^2 \int_{|\varepsilon k|<\delta} (1+|k|)^{2\ell} |\Sigma e^{\frac{t}{\varepsilon}\alpha_j(\varepsilon k)} (e_j(-\varepsilon k) - e_j(0), \hat{f})_{L^2(v)} e_j(\varepsilon k)|^2_{L^2(v)} dk \\ &\le C^2 \int_{|\varepsilon k|<\delta} \varepsilon^2 k^2 e^{-t\varepsilon\alpha_0 k^2} |\hat{f}(k,\cdot)|^2_{L^2(v)} (1+|k|)^{2\ell} dk \end{aligned}$$

$$\leq C^2\Big(\int_{|k|<\delta} (\quad)dk + \int_{\delta<|k|<\delta/\varepsilon} (\quad)dk\Big)$$

$$\leq C^2\big(\varepsilon^2 \| f\|^2_{L^{2,1}} / (1+\varepsilon t)^{5/2} + \delta^2 \| f\|^2_{\ell}\, e^{-\varepsilon t \alpha_0 \delta^2}\big) .$$

qed.

Before we treat the nonlinear Boltzmann equation we improve the decay estimates in theorem 3.1 into those in the space of $B_{m,\ell}$ $(m\geq 3,\ \ell\geq 2)$.

Definition 3.1. $\mathring{\mathcal{C}}([0,\infty);X)$ denotes the space of functions $f(t)$ which is continuous of $t \in [0,\infty)$ with the values in the Banach space X and which decays to zero in X as $t \to \infty$. The norm of $f(t) \in \mathring{\mathcal{C}}([0,\infty);X)$, where $X = H_\ell$ or $B_{m,\ell}$, $(\ell,\ m\geq 0)$, is defined by

$$(3.8)\qquad \begin{aligned} |||\, f(\cdot)\, |||_\ell &= \max_{0\leq t<\infty} \| f(t)\|_\ell \\ |||\, f(\cdot)\, |||_{m,\ell} &= \max_{0\leq t<\infty} \| f(t)\|_{m,\ell} \end{aligned}$$

Theorem 3.2. The linear Boltzmann operator $\frac{1}{\varepsilon} B_\varepsilon$ generates the strongly continuous semigroup $e^{\frac{t}{\varepsilon}B_\varepsilon}$ also in $B_{m,\ell}$ $(m\geq 2,\ell\geq 0)$. Let $f \in B_{m,\ell}$. Then we have a constant $C_1 < +\infty$ such that for $f(t) = e^{\frac{t}{\varepsilon}B_\varepsilon} f$

$$(3.9)\qquad \begin{aligned} &\| f(t)\|_{m,\ell} \leq C_1 \| f\|_{m,\ell} \quad \text{and} \\ &\| f(t)\|_{m,\ell} \to 0 \quad \text{as} \quad t \to \infty . \end{aligned}$$

Moreover define

$$(3.10)\qquad h(t) = \int_0^t e^{\frac{t-s}{\varepsilon}B_\varepsilon} \frac{1}{\varepsilon} \nu\Gamma(g(s),\ g(s))ds,$$

where $g(t) \in \mathring{\mathcal{C}}([0,\infty);B_{m,\ell})$ for some $m\geq 3$, $\ell\geq 2$. Then $h(t) \in \mathring{\mathcal{C}}([0,\infty);B_{m,\ell})$ and we have a constant $C_2 < \infty$ such that

$$|||\, h(\cdot)\, |||_{m,\ell} \le \frac{C_2}{\varepsilon^2} \; |||\, g(\cdot)\, |||^{2}_{m,\ell}$$

<u>Proof</u> Following Grad (1965) we use the representation

$$f(t) = e^{\frac{t}{\varepsilon}B_\varepsilon} f + \int_0^t e^{\frac{t-s}{\varepsilon}B_\varepsilon} \frac{1}{\varepsilon}\, \nu\Gamma(g(s),g(s))ds$$

$$(3.11) \qquad = e^{\frac{t}{\varepsilon}A_\varepsilon} f + \int_0^t e^{\frac{t-s}{\varepsilon}A_\varepsilon} \frac{1}{\varepsilon}\, \nu\Gamma(g(s),g(s))ds \; +$$

$$+ \int_0^t e^{\frac{t-s}{\varepsilon}A_\varepsilon} \frac{1}{\varepsilon}\, Kf(s)ds \; .$$

$e^{\frac{t}{\varepsilon}B_\varepsilon}$ is a strongly continuous semigroup in $B_{m,\ell}$ $(m\ge 2,\ \ell\ge 0)$ because of the definition of the space $B_{m,\ell}$ with (2.6) and by (3.11) with $g \equiv 0$ and (2.18). The decay (3.9) follows from (3.11) with $g \equiv 0$ and from (2.18) :

$$\|\, f(t)\,\|_{0,\ell} \le e^{-\nu_0 t/\varepsilon} \|\, f\,\|_{0,\ell} + \|\int_0^t e^{\frac{t-s}{\varepsilon}A_\varepsilon} \frac{1}{\varepsilon}\, Kf(s)ds\|_{0,\ell}$$

$$(3.12) \qquad \le e^{-\nu_0 t/\varepsilon} \|\, f\,\|_{0,\ell} + C\left[\int_0^{t/2} + \int_{t/2}^{t}\right] \frac{e^{-\nu_0 (t-s)/\varepsilon}}{\varepsilon} \|\, f(s)\,\|_{\ell}\, ds$$

$$\to 0 \quad \text{as} \quad t \to \infty \; ,$$

where (3.5) is used. Successively for $j=1,2,\cdots,m$

$$\|\, f(t)\,\|_{j,\ell} \le e^{-\nu_0 t/\varepsilon} \|\, f\,\|_{j,\ell} + C\left[\int_0^{t/2} + \right.$$

$$(3.13) \qquad \left. + \int_{t/2}^{t}\right] \frac{e^{-\nu_0 (t-s)/\varepsilon}}{\varepsilon} \|\, f(s)\,\|_{j-1,\ell}\, ds$$

$$\to 0 \quad \text{as} \quad t \to \infty \; .$$

For the latter half of the theorem we note that if $g(s)\in B_{m,\ell}$ for some $m > 5/2$, $\ell > 3/2$, then by lemma 2.2.

$$\nu\Gamma(g(s),g(s)) \in H_{m-1,\ell} \cap L^2(v;L^1(x)) \subset H_\ell \cap L^2(v;L^1(x))$$

and

$$(\nu\Gamma(g(s),g(s)),\ \Psi_j)_{L^2(v)} = 0\ ,\quad j=1,\cdots,5\ .$$

Thus the rapid decay estimate (ii) in theorem 3.1 applies to this case and we have

$$\| h(t) \|_{\ell} \le \int_0^t \frac{C(\| \nu\Gamma(g(s),g(s)) \|_{\ell} + \| \nu\Gamma(g(s),g(s)) \|_{L^{2,1}})}{\varepsilon(1+\varepsilon(t-s))^{5/4}} ds$$

$$\le C \int_0^t \frac{\| g(s) \|^2_{m,\ell}}{\varepsilon(1+\varepsilon(t-s))^{5/4}} ds \le C \int_0^{t/2} + \int_{t/2}^t ds$$

$$= \frac{C(\max_{0\le s\le t/2} \| g(s) \|_{m,\ell})^2}{\varepsilon^2(1+\varepsilon t/2)^{1/4}} + \frac{C}{\varepsilon^2}(\max_{t/2\le s\le t} \| g(s) \|_{m,\ell})^2$$

$$\le \frac{C}{\varepsilon^2}(||| g(\cdot) |||_{m,\ell})^2 \quad \text{and tends to zero as} \quad t \to +\infty.$$

To get the decay of $h(t)$ in $B_{m,\ell}$ we use (3.11) with $f = 0$ in the same way as (3.12) (3.13).

$$\| h(t) \|_{j,\ell} \le \sup_v \int_0^t e^{-\frac{t-s}{\varepsilon}\nu(v)} \frac{1}{\varepsilon} \nu(v) \| \Gamma(g(s),g(s)) \|_{j,\ell} ds$$

$$+ \int_0^t e^{-\frac{t-s}{\varepsilon}\nu_0} \frac{1}{\varepsilon} C \| h(s) \|_{j-1,\ell} ds$$

$$\le C\{ \max_{0\le s\le t/2} \| g(s) \|^2_{m,\ell} \cdot \frac{e^{-\nu_0 t/2\varepsilon}}{\varepsilon} + \max_{t/2\le s\le t} \| g(s) \|^2_{m,\ell}$$

$$+ (\max_{0\le s\le t/2} \| h(s) \|_{j-1,\ell}) \frac{e^{-\nu_0 t/2\varepsilon}}{\varepsilon} + \max_{t/2\le s\le t} \| h(s) \|_{j-1,\ell}\}$$

for $j=m,m-1,\cdots,2,1$ and also

$$\| h(t) \|_{0,\ell} \le C\{ \max_{0\le s\le t/2} \| g(s) \|^2_{m,\ell} \frac{e^{-\nu_0 t/2\varepsilon}}{\varepsilon} + \max_{t/2\le s\le t} \| g(s) \|^2_{m,\ell} \}$$

$$\le \frac{C_2}{\varepsilon^2} \| g(\cdot) \|^2_{m,\ell}$$

$$\to 0 \quad \text{as} \quad t \to \infty\ .$$

qed.

Now we consider the nonlinear Boltzmann equation

(3.14) $\frac{\partial f(t)}{\partial t} = \frac{1}{\varepsilon} B_\varepsilon f(t) + \frac{1}{\varepsilon} \nu\Gamma(f(t),f(t))$ in $t \geq 0$,

in the space $B_{m,\ell}$ $(m\geq 3,\ \ell\geq 2)$ with the initial condition

(3.15) $f(0) = f(x,v) \in B_{m,\ell}$ $(m\geq 3,\ \ell\geq 2)$.

The solution is constructed by the successive approximation $(n=0,1,2,\cdots)$

(3.16) $f^{(n)}(t)=e^{\frac{t}{\varepsilon}B_\varepsilon}f + \int_0^t e^{\frac{t-s}{\varepsilon}B_\varepsilon}\frac{1}{\varepsilon}\nu\Gamma(f^{(n-1)}(s),f^{(n-1)}(s))ds$

in the Banach space $\mathring{\mathcal{C}}([0,+\infty);B_{m,\ell})$, $f^{(-1)}(t) \equiv 0$. Let the initial data $f(0)$ have $E = \| f(0) \|_{m,\ell} < \infty$ for some $m \geq 3$, $\ell \geq 2$. Then by theorem 3.2 and by (3.16) we have for the same m, ℓ

$$f^{(n)}(t) \in \mathring{\mathcal{C}}([0,\infty);B_{m,\ell}) \quad \text{and}$$

$$|||f^{(n)}(\cdot)|||_{m,\ell} \leq C_1 E + \frac{C_2}{\varepsilon^2}(|||f^{(n-1)}(\cdot)|||_{m,\ell})^2 ,$$

$$|||f^{(n+1)}(\cdot)-f^{(n)}(\cdot)|||_{m,\ell} \leq \frac{C_2}{\varepsilon^2}(|||f^{(n)}(\cdot)|||_{m,\ell} + |||f^{(n-1)}(\cdot)|||_{m,\ell})\,|||f^{(n)}(\cdot)-f^{(n-1)}(\cdot)|||_{m,\ell} .$$

for $n=0,1,2,\cdots$.

Therefore if we suppose $0 \leq E < \varepsilon^2/4\,C_1\,C_2$ and set $a = 1 - \sqrt{1-4\,C_1\,C_2\,E/\varepsilon^2} < 1$, we get

$$|||f^{(n)}(\cdot)|||_{m,\ell} \leq a/2\,C_2 \quad \text{and}$$

$$|||f^{(n+1)}(\cdot) - f^{(n)}(\cdot)|||_{m,\ell} \leq a|||f^{(n)}(\cdot) - f^{(n-1)}(\cdot)|||_{m,\ell} .$$

Then $f^{(n)}(t)$ converges in $\mathring{\mathcal{C}}([0,\infty);B_{m,\ell})$ to $f(t)$, which is a unique solution of (3.14) (3.15) and decays to zero in $B_{m,\ell}$ as $t \to \infty$.

Theorem 3.3. Let the initial data $f \in B_{m,\ell}$ for some $m \geq 3$, $\ell \geq 2$. Then there exists a constant $E_0 > 0$ such that if $E = \| f \|_{m,\ell} < \varepsilon^2 E_0$, the solution $f(t)$ of Boltzmann equation (3.14) (3.15) with ε fixed exists in the space $B_{m,\ell}$ uniquely in the large in time and decays to zero as $t \to \infty$.

Remark 3.1.

(i) Theorem 3.3. means that the solution to the initial value problem for Boltzmann equation (1.1) converges to the absolute Maxwellian distribution as $t \to \infty$, provided that the initial deviation from it is small in the norm of $B_{m,\ell}$ ($m \geq 3$, $\ell \geq 2$).

(ii) If $m \geq 3$ and $\ell \geq 3$, the solution is smooth and satisfies Boltzmann equation in the classical sense.

(iii) The uniqueness of the solution is just proved in a small (in the norm of $B_{m,\ell}$ ($m \geq 3$, $\ell \geq 2$)) neighbourhood of the absolute Maxwellian distribution cf. Shizuta (preprint).

4. THE FLUID DYNAMICAL LIMIT OF BOLTZMANN EQUATION AT THE LEVEL OF COMPRESSIBLE EULER EQUATION

Let us consider the initial value problem for Boltzmann equation with $\varepsilon \in (0,1]$

$$(4.1) \quad \frac{\partial F_\varepsilon(t)}{\partial t} = -\Sigma v_j \frac{\partial F_\varepsilon(t)}{\partial x_j} + \frac{1}{\varepsilon} Q(F_\varepsilon(t), F_\varepsilon(t)) \quad \text{in} \quad t \geq 0 ,$$

$$(4.2) \qquad F_\varepsilon(0) = F(x,v) \geq 0 .$$

First we note the non-negativity of the solution $F_\varepsilon(t,x,v)$ for fixed $\varepsilon \in (0,1]$.

Theorem 4.1. Let $F(x,v) = M(v)+M(v)^{1/2} f(x,v) \geq 0$ and $f(x,v) \in B_{m,\ell}$ for some $m \geq 3$, $\ell \geq 2$. Then there exist two constants $E_0 > 0$ and $t_0 > 0$ such that if $\| f \|_{m,\ell} < E_0$, then there exists a unique non-negative solution to (4.1) (4.2) in $0 \le t \le \varepsilon t_0$.

The solution is given by the iteration which preserves the non-negativity.

$$(4.3) \quad \frac{\partial F^{(n+1)}(t)}{\partial t} + v \cdot \frac{\partial F^{(n+1)}(t)}{\partial x} = \frac{1}{\varepsilon} \int (F_*^{(n)'} F^{(n)'} - F_*^{(n)} F^{(n+1)}) d\omega$$

where $d\omega = Vr\, dr\, d\phi\, dv_*$,

$$(4.4) \quad F^{(n+1)}(0) = F(x,v) , \quad n=0,1,2,\cdots \quad \text{and}$$

$$(4.5) \quad F^{(0)}(t) = F(x,v) \ge 0 .$$

The proof of the convergence of the iteration uses a modified argument of Grad (1965). By the uniqueness of solutions near to the absolute Maxwellian for problem (4.1) (4.2) the solution as the limit of $n \to \infty$ coincides to the solution given by Grad (1965). cf. Nishida (preprint)

We seek the solution of (4.1) (4.2) in $0 \le t < t_0$, where t_0 is independent of $\varepsilon \in (0,1]$, again around the absolute Maxwellian distribution, i.e., of the integral equation

$$(4.6) \quad f_\varepsilon(t) = e^{\frac{t}{\varepsilon}B_\varepsilon} f(0) + \int_0^t e^{\frac{t-s}{\varepsilon}B_\varepsilon} \frac{1}{\varepsilon} \nu\Gamma(f_\varepsilon(s), f_\varepsilon(s))ds$$

$$\text{for} \quad \varepsilon \in (0,1] .$$

Let $f(0) \in B_{m,\ell,\rho_0}$ for some $m \ge 3$, $\ell \ge 2$, $\rho_0 > 0$. The solution of (4.6) is sought in the Banach space B, which is defined by

Definition 4.1.

B = {f(t) ; continuous function of t with the values in $B_{m,\ell,\rho}$, which has the norm

$$(4.7)\quad N_a[f] = \sup_{\substack{0\le\rho<\rho_0\\ 0\le t<a(\rho_0-\rho)}} |||f(t)|||_{m,\ell,\rho}(1-t/a(\rho_0-\rho)) < \infty$$

for a suitable small $a > 0\}$.

Theorem 4.2. Let the initial data have the norm

$$(4.8)\quad E = |||f(0)|||_{m,\ell,\rho_0} < +\infty \quad \text{for some} \quad m\ge 3,\ \ell\ge 2,\ \rho_0>0 .$$

Then there exists $E_1 > 0$, $a > 0$ and $C_1 < \infty$ such that for any $f(0)$ with $E < E_1$ and for any $\varepsilon \in (0,1]$ the equation (4.6) has the unique solution $f_\varepsilon(t)$, which is continuous of $t, 0 \le t < a(\rho_0-\rho)$ with the values in $B_{m,\ell,\rho}$, $0 < \rho < \rho_0$ and has the uniform bounds

$$(4.9)\quad |||f_\varepsilon(t)|||_{m,\ell,\rho_0} \le C_1E \quad \text{in} \quad 0\le t<a(\rho_0-\rho),\ 0\le\rho<\rho_0 ,$$

where C_1 is independent of $\varepsilon \in (0,1]$.

The proof of theorem 4.2 is based on the following proposition.

Proposition 4.1. The solution of linear Boltzmann equation has a uniform estimate :

$$(4.10)\quad |||e^{\frac{t}{\varepsilon}B_\varepsilon}f(0)|||_{m,\ell,\rho_0} \le C|||f(0)|||_{m,\ell,\rho_0}$$

for $m \ge 3$, $\ell \ge 2$, $\rho_0 > 0$,

where C is independent of $\varepsilon \in (0,1]$. Furthermore let us consider the function for any $f(t),g(t) \in B$, $m \ge 3$, $\ell \ge 2$,

$$(4.11)\quad h(t) = \int_0^t e^{\frac{t-s}{\varepsilon}B_\varepsilon}\frac{1}{\varepsilon}\nu\Gamma(f(s),g(s))ds .$$

Then it has a uniform estimate

$$(4.12)\quad N_b[h] \le CRN_b[g] \le CRN_a[g] \quad \text{for any} \quad b < a ,$$

where $N_b[h]$ is defined by (4.7) with b replacing a and

$$(4.13)\quad R = \sup_{\substack{0\le s<b(\rho_0-\rho)\\ 0\le\rho<\rho_0}} |||f(s)|||_{m,\ell,\rho} .$$

Proof of Proposition 4.1.

By theorem 2.1 (2.27) and Planchrel theorem we have

$$|||e^{\frac{t}{\varepsilon}B_\varepsilon}f(0)|||^2_{\ell,\rho_0} = \int |e^{\frac{t}{\varepsilon}B_{\varepsilon k}} e^{|k|\rho_0}(1+|k|)^\ell \hat{f}(0,k,v)|^2_{L^2(v)} dk$$

$$\le |||f(0)|||^2_{\ell,\rho_0} \quad \text{for any } \ell \ge 0,\ \rho_0 > 0 .$$

It is improved to the estimate in the norm of B_{m,ℓ,ρ_0}, $m\ge 3$, if we remember the representation

$$(4.14)\quad e^{\frac{t}{\varepsilon}B_\varepsilon}f(0) = e^{\frac{t}{\varepsilon}A_\varepsilon}f(0) + \int_0^t e^{\frac{t-s}{\varepsilon}A_\varepsilon}\frac{K}{\varepsilon}(e^{\frac{s}{\varepsilon}A_\varepsilon}f(0))ds$$

and the same argument used in the proof of (3.9).

The latter half of the proposition is proved as follows : since $(e_j(0),\nu\Gamma(f,g))_{L^2(v)} = 0$, $j=1,2,\cdots,5$, we have by theorem 2.1

$$h(t) = \int_0^t [\frac{1}{(2\pi)^{3/2}} \int_{|\varepsilon k|<\delta} \{ \sum_{j=1}^5 e^{\frac{t-s}{\varepsilon}\alpha_j(\varepsilon k)} ik(e'_j(-\theta\varepsilon k),(\nu\Gamma)^\wedge)e_j(\varepsilon k)$$

$$+ e^{\frac{t-s}{\varepsilon}A_{\varepsilon k}}\frac{1}{\varepsilon}(\nu\Gamma)^\wedge + e^{-\frac{t-s}{\varepsilon}\beta_1}\frac{1}{\varepsilon}Z_1(\varepsilon k,t/\varepsilon)(\nu\Gamma)^\wedge\}dk$$

$$+ \int_{|\varepsilon k|>\delta} \{e^{\frac{t-s}{\varepsilon}A_{\varepsilon k}}\frac{1}{\varepsilon}(\nu\Gamma)^\wedge + e^{-\frac{t-s}{\varepsilon}\beta_2}\frac{1}{\varepsilon}Z_2(\varepsilon k,t/\varepsilon)(\nu\Gamma)^\wedge\}dk]ds .$$

The norm in $H_{\ell,\rho}$ has the estimate by the same theorem

$$|||h(t)|||_{\ell,\rho} \le C\int_0^t [(\int (1+|k|)^{2\ell} e^{2|k|\rho}k^2|(\nu\Gamma)^\wedge(s)|^2_{L^2(v)} dk)^{1/2}$$

$$+ \frac{e^{-\frac{t-s}{\varepsilon}\beta_0}}{\varepsilon} |||\nu\Gamma(s)|||_{\ell,\rho}]ds$$

$$\leq C\{\int_0^t \frac{|||f(s)|||_{m,\ell,\rho(s)}|||g(s)|||_{m,\ell,\rho(s)}}{\rho(s)-\rho}ds$$

$$+\int_0^t \frac{e^{-\frac{t-s}{\varepsilon}\beta_0}}{\varepsilon}|||f(s)|||_{m,\ell,\rho}|||g(s)|||_{m,\ell,\rho}\,ds$$

for some choice of $\rho(s)$, $\rho < \rho(s) < \rho_0 - s/a$, where we used lemmas 2.1 and 2.2.

It can be estimated by (4.13) in $0 \leq t < b(\rho_0-\rho)$, $0 \leq \rho < \rho_0$ for any $b < a$

$$|||h(t)|||_{\ell,\rho} \leq CR(\int_0^t \frac{|||g(s)|||_{m,\ell,\rho(s)}}{\rho(s)-\rho}ds$$

$$+\int_0^t \frac{e^{-\frac{t-s}{\varepsilon}\beta_0}}{\varepsilon}|||g(s)|||_{m,\ell,\rho}\,ds)$$

$$\leq CRN_b[g](\int_0^t \frac{ds}{(\rho(s)-\rho)(1-s/b(\rho_0-\rho(s)))}$$

$$+\int_0^t \frac{e^{-\frac{t-s}{\varepsilon}\beta_0}}{\varepsilon}\frac{ds}{1-s/b(\rho_0-\rho)})$$

with $\rho < \rho(s) < \rho_0 - s/b$.

Therefore if we choose $\rho(s) = (\rho_0 - s/b + \rho)/2$, we have

$$(4.15)\quad \sup_{\substack{0\leq\rho<\rho_0\\ 0\leq t<b(\rho_0-\rho)}} |||h(t)|||_{\ell,\rho}(1-t/b(\rho_0-\rho)) \leq C(4b+1/\beta_0)RN_b[g]$$

In order to obtain the estimate for $N_b[h]$ from (4.15) we use the equivalent representation

$$(4.16)\quad h(t) = \int_0^t e^{\frac{t-s}{\varepsilon}A_\varepsilon}\frac{1}{\varepsilon}\nu\Gamma(f(s),g(s))ds + \int_0^t e^{\frac{t-s}{\varepsilon}A_\varepsilon}\frac{K}{\varepsilon}h(s)ds$$

and the same argument as that for (3.10). Thus we arrive at

$$N_b[h] \leq CRN_b[g] \leq CRN_a[g] .$$

qed of proposition 4.1.

Now we introduce the same approximation as (3.16) to solve (4.6), i.e.,

$$f^{(0)}(t) = e^{\frac{t}{\varepsilon}B_\varepsilon} f(0)$$

$$g^{(0)}(t) = \int_0^t e^{\frac{t-s}{\varepsilon}B_\varepsilon} \frac{1}{\varepsilon} \nu\Gamma(f^{(0)}(s), f^{(0)}(s))ds ,$$

$$f^{(1)}(t) = g^{(0)}(t) + f^{(0)}(t) ,$$

$$g^{(n)}(t) = \int_0^t e^{\frac{t-s}{\varepsilon}B_\varepsilon} \frac{1}{\varepsilon} \{\nu\Gamma(f^{(n)}(s), g^{(n-1)}(s)) + \nu\Gamma(g^{(n-1)}(s), f^{(n-1)}(s))\}ds ,$$

$$f^{(n+1)}(t) = g^{(n)}(t) + f^{(n)}(t) = f^{(0)}(t) + \int_0^t e^{\frac{t-s}{\varepsilon}B_\varepsilon} \frac{1}{\varepsilon} \nu\Gamma(f^{(n)}(s), f^{(n)}(s))ds ,$$

$$n = 1,2,\cdots .$$

It is easy from proposition 4.1 to see that for any $\rho \le \rho_0$

$$(4.18) \quad |||f^{(0)}(t)|||_{m,\ell,\rho} \le C|||f(0)|||_{m,\ell,\rho} \le C|||f(0)|||_{m,\ell,\rho_0} \equiv R_0$$

and that

$$(4.19) \quad \mu_0 \equiv \sup_{\substack{0\le\rho<\rho_0 \\ 0\le t<a_0(\rho_0-\rho)}} |||g^{(0)}(t)|||_{m,\ell,\rho} \le CR_0^2$$

$$\text{for any } a_0 > 0 .$$

Then it follows from (4.17) and (4.19) that

$$(4.20) \quad |||f^{(1)}(t)|||_{m,\ell,\rho} \le R_0 + \mu_0$$

$$\text{in } 0 \le \rho < \rho_0,\ 0 \le t < a_0(\rho_0-\rho).$$

Define $a_1 = a_0 > 0$ and

$$(4.21) \quad a_{n+1} = a_n - \frac{a_0}{2^{n+1}} \quad \text{for } n = 1,2,\cdots$$

and

(4.22) $\quad N_n[g] = N_{a_n}[g] \qquad$ for $\quad n = 0,1,2,\cdots$.

By use of proposition 4.1 and by the induction we have for $k = 1,2,\cdots$

(4.23) $\quad \mu_k \equiv N_k[g^{(k)}] \le CR\mu_{k-1} \le \mu_0/3^k$,

provided

(4.24) $\quad R < 1/3C$,

and also

$$|||f^{(k+1)}(t)|||_{m,\ell,\rho} \le |||g^{(k)}(t)|||_{m,\ell,\rho} + |||f^{(k)}(t)|||_{m,\ell,\rho}$$

$$\le \frac{\mu_k}{1-\dfrac{t}{a_k(\rho_0-\rho)}} + |||f^{(k)}(t)|||_{m,\ell,\rho}$$

$$\le \frac{a_k\mu_k}{a_k-a_{k+1}} + |||f^{(k)}(t)|||_{m,\ell,\rho}$$

$$\le 2\sum_{j=1}^{\infty}\left(\frac{2}{3}\right)^j\mu_0 + \mu_0 + |||f^{(0)}(t)|||_{m,\ell,\rho}$$

(4.25) $$\le 5CR_0^2 + R_0 < R$$

in $0 \le t < a_{k+1}(\rho_0-\rho)$, $0 \le \rho < \rho_0$,

provided that R_0 is small. Thus if we choose R_0 small, (4.24) and (4.25) are valid. Therefore there exists

$$\lim_{k\to\infty} f^{(k+1)}(t) = f(t) ,$$

the limit of which is the solution of (4.6) for each $\varepsilon \in (0,1]$ and has the uniform bounds by (4.25)

(4.26) $\quad |||f(t)|||_{m,\ell,\rho} \le R \qquad$ in $\quad 0 \le t < a(\rho_0-\rho)$, where R and $a = \lim_{n\to\infty} a_n$ are independent of $\varepsilon \in (0,1]$.

qed of theorem 4.2

In order to take the limit of $f_\varepsilon(t)$ as $\varepsilon \to 0$ we need more than the uniform bounds (4.9). The uniform continuity in t is given by the following.

Theorem 4.3. Let the initial data $f(0) \in B_{m,\ell,\rho_0}$ for some $m \geq 3$, $\ell \geq 2$, $\rho_0 > 0$ and let

$$E = |||f(0)|||_{m,\ell,\rho_0} < E_1 ,$$

where E_1 is defined in theorem 4.2. Then there exist constants $0 < E_2 \leq E_1$ and $C_2 < \infty$ such that if $E < E_2$, then the solution $f_\varepsilon(t)$ of (4.6) has the uniform Hölder-continuity in t :

$$(4.27)\quad |||f_\varepsilon(t)-f_\varepsilon(s)|||_{m-\sigma,\ell-\sigma,\rho} \leq C_2 E\left\{ \frac{t-s}{s(1-t/a(\rho_0-\rho))} \right\}^\sigma$$

for $0 < s < t$ and for a fixed $\sigma \in (0,1/2)$,

where C_2 is independent of $\varepsilon \in (0,1]$.

The proof needs the Hölder-continuity of the solution for the linear Boltzmann equation and of the function h(t) (4.11). See Nishida (preprint). We only remark that the Hölder-coefficient has the singularity of $1/t^\sigma$ as $t \to 0$, which corresponds to the initial layer of the rarefied gas motion described by Boltzmann equation.

It follows from theorems 4.2 and 4.3 that by Ascoli-Arzela lemma we can choose a convergent subsequence as $\varepsilon \to 0$ such that

$$(4.28)\quad f_\varepsilon(t) \to f_0(t) \quad \text{in} \quad B_{m-\sigma,\ell-\sigma,\rho} , \quad 0\leq t<a(\rho_0-\rho),\ 0\leq\rho<\rho_0 .$$

The limit function has the bound

(4.29) $|||f_0(t)|||_{m,\ell,\rho} \le C_1 E$ in $0 \le t < a(\rho_0-\rho)$, $0 \le \rho < \rho_0$.

and the Hölder-continuity of $\sigma \in (0,1/2)$.

(4.30) $|||f_0(t)-f_0(s)|||_{m-\sigma,\ell-\sigma,\rho} \le C_2 E\{ \frac{t - s}{s(1-t/a(\rho_0-\rho))} \}^\sigma$.

Now we turn to the original mass density distribution function

(4.31) $F_\varepsilon(t,x,v) = M(v) + M(v)^{1/2} f_\varepsilon(t,x,v)$,

which satisfies Boltzmann equation (4.1) (4.2). Taking the limit of the equation (4.1) in the integrated form in t along the subsequence (4.28) as $\varepsilon \to 0$, we have by the uniform bound (4.29)

(4.32) $Q(F_0(t,x,v),F_0(t,x,v)) = 0$ in $0 < t < a\rho_0$,

where $F_0(t,x,v) = M(v) + M(v)^{1/2} f_0(t,x,v)$.
If we assume that $F(0,x,v) = F(x,v) \ge 0$ and $\rho(0,x) = \int F(x,v) dv > 0$ in $x \in R^3$, the solution has the same properties by theorem (4.1) and by the mass conservation laws (1.10) :

(4.33) $F_0(t,x,v) \ge 0$

(4.34) $\rho(t,x) = \int F_0(t,x,v) dv > 0$.

It follows from (4.32) (4.33) (4.34) that $F_0(t,x,v) > 0$ and then $F_0(t,x,v)$ is locally Maxwellian. Thus we can obtain the conservation laws (1.13) for $F_0(t,x,v)$ from (1.10) for $F_\varepsilon(t, x,v)$ as the limit of $\varepsilon \to 0$ along the subsequence of (4.28). The uniqueness of the solution to the initial value problem (1.13) guarantees the convergence of full sequence F_ε to F_0 as $\varepsilon \to 0$.

Theorem Let the initial data $F(x,v) = M(v)+M(v)^{1/2}f(x,v) \geq 0$ with $\rho(0,x) = \int F(x,v)dv > 0$ in $x \in R^3$, and let $f(x,v) \in B_{m,\ell,\rho_0}$ for some $m \geq 3$, $\ell \geq 2$, $\rho_0 > 0$ and set $|||f|||_{m,\ell,\rho_0} = E$. If $E < E_2$, where E_2 is defined in Theorem 4.3., then the solution $F_\varepsilon(t,x,v)$ of Boltzmann equation (4.1) (4.2) exists uniquely in $B_{m,\ell,\rho}$, $0 \leq t < a(\rho_0-\rho)$, $0 \leq \rho < \rho_0$ for any $\varepsilon \in (0,1]$ and is non-negative there, where a is defined in Theorem 4.2. Furthermore there exists

$$\lim_{\varepsilon\to 0} F_\varepsilon(t,x,v) = F_0(t,x,v) \quad \text{in} \quad B_{m,\ell,\rho}\ , \quad 0<t<a(\rho_0-\rho),\ 0<\rho<\rho_0\ ,$$

where $F_0(t,x,v)$ is locally Maxwellian distribution. Therefore its fluid dynamical quantities satisfy the conservation laws (1.13).

At last we note that the system (1.13) with (1.11) is hyperbolic and has two genuinely nonlinear characteristic fields, and so it developes in general shock waves in finite time even for the analytic initial data.

REFERENCES

1. A. Arseniev (1965), The Cauchy problem for the linearized Boltzmann equation, J. Comp. Math. & Math. Phys. (USSR), 5, 864-882.

2. S. Chapman and T. G. Cowling (1970), The Mathematical Theory of Non-Uniform Gases, Cambridge Univ. Press, III edition, London.

3. R. Ellis and M. Pinsky (1975), The first and second fluid approximations to the linearized Boltzmann equation, J. Math. Pures Appl., 54, 125-156.

4. ________ and ________ (1975), The projection of the Navier-Stokes equations upon the Euler equation, J. Math. Pures Appl., 54, 157-181.

5. H. Grad (1958), Principles of the kinetic theory of gases, Handbuch der Physik, XII, 205-294, Springer-Verlag,

Berlin.

6. _______ (1963), Asymptotic theory of the Boltzmann equation, Phys. of Fluids, 6, 147-181.

7. _______ (1963), Asymptotic theory of the Boltzmann equation II, Rarefied Gas Dynamics, I, 26-59, ed. by J. Laurmann, Academic Press.

8. _______ (1965), Asymptotic equivalence of the Navier-Stokes and nonlinear Boltzmann equations, Proc. Symp. in Appl. Math., Amer. Math. Soc., 17, 154-183.

9. _______ (1965), On Boltzmann's H-theorem, J. Soc. Indust. Appl. Math., 13, 259-277.

10. _______ (1969), Singular and nonuniform limits of solutions of the Boltzmann equation, in Transport Theory, Proc. Symp. Appl. Math. of AMS-SIAM, New York, 1967, Amer. Math. Soc., 269-308.

11. J. Guiraud (1973), The Boltzmann equation in kinetic theory, A survey of mathematical results, XIth Symp. on Advanced Prob. and Methods in Fluid Mechanics, Warszawa.

12. _________ (1974), An H.theorem for a gas of rigid spheres in a bounded domain, Coll. Intern. CNRS, Théories Cinétiques Classiques et Relativistes, Paris.

13. D. Hilbert (1912), Grundzüge einer allgemeinen Theorie der linearen Integralgleichungen, Leipzig-Berlin.

14. J. McLennan (1965), Convergence of the Chapman-Enskog expansion for the linearized Boltzmann equation, Phys. of Fluids, 8, 1580-1584.

15. B. Nicolaenko and J. Thurber (1975), Weak shock and bifurcating solutions of the nonlinear Boltzmann equation, J. de Mécanique, 14, 305-338.

16. B. Nicolaenko (1974), Shock wave solutions of the Boltzmann equation as a nonlinear bifurcation problem from the essential spectrum, Coll. Intern. CNRS n° 236, Théories Cinétiques Classiques et Relativistes, 127-150, Paris.

17. ___________ (1976), A general class of nonlinear bifurcation problems from a point in the essential spectrum ; Appl. to shock wave solutions of kinetic equations, Symp. on Appl. of Bifurcation Theory, Univ. of Wisconsin-Madison.

18. L. Nirenberg (1972), An abstract form of the nonlinear Cauchy-Kowalewski theorem, J. Diff. Geometry, 6, 561-576.

19. T. Nishida and K. Imai (1976), Global solutions to the initial value problem for the nonlinear Boltzmann equation, Publ. Res. Inst. Math. Sci. Kyoto Univ., 12, 229-239.

20. T. Nishida, Fluid dynamical limit of the nonlinear Boltzmann equation at the level of the compressible Euler equation, to appear in Comm. Math. Phys.

21. __________, On the Nirenberg's abstract form of the nonlinear Cauchy-Kowalewski theorem, to appear in J. Diff. Geometry.

22. Y. Pao (1974), Boltzmann collision operator with inverse-power intermolecular potentials, I, Comm. Pure Appl. Math., 27, 407-428.

23. ______ (1974), Boltzmann collision operator with inverse-power intermolecular potentials, II, Comm. Pure Appl. Math., 27, 559-581.

24. M. Pinsky (1976), On the Navier-Stokes approximation to the linearized Boltzmann equation, J. Math. Pure Appl., 55, 217-231.

25. G. Scharf (1969), Normal solutions of the linearized Boltzmann equation, Helv. Phys. Acta, 42, 5-22.

26. Y. Shizuta, On the classical solutions of the Boltzmann equation, to appear in Comm. Pure Appl. Math.

27. __________, The existence and approach to equilibrum of classical solutions of the Boltzmann equation, to appear in Comm. Math. Phys.

28. __________ and K. Asano (1977), Global solutions of the Boltzmann equation in a bounded convex domain, Proc. Japan Acad., 53, No. 1, Ser. A, 3-5.

29. S. Ukai (1974), On the existence of global solutions of mixed problem for nonlinear Boltzmann equation, Proc. Japan Acad., 50, 179-184.

30. _______ (1976), Les solutions globales de l'équation nonlinéaire de Boltzmann dans l'espace tout entier et dans le demi-espace, Compte Rendu Acad. Sci. Paris, 282, Série A, 317-320.

31. _______ and T. Nishida, On the Boltzmann equation, to appear in Proc. Coll. Franco-Japon at Tokyo and Kyoto (1976), Springer-Verlag lecture note.

Department of Applied Mathematics and Physics
Kyoto University
Kyoto 606, Japan

On Some Problems Connected with Navier Stokes Equations

J. L. Lions

INTRODUCTION.

We give in this lecture some results on the flows of non homogeneous fluids (Section 1) and on the flows of fluids in media with "many" obstacles in a periodic structure (Section 2).

In Section 1 we use a simple model, which, although an oversimplification of "real" problems, is of some physical interest and leads to very many interesting mathematical (and numerical[1]) questions. We consider Newtonian fluids in Sections 1.1 to 1.5, non Newtonian fluids of Bingham's type being briefly considered in Section 1.6.1 ; the fluids are non homogeneous; we use a semi-Galerkin's approximation (i.e. a Galerkin's approximation on the velocity, and an infinite dimensional approximation on the density) and we introduce a priori estimates of Kajikov [12]; these a priori estimates give results on some sort of fractional derivative in time of the velocity (they are weaker than those obtained by Fourier transform in t , as in Lions [23], but the Fourier transform method seems to fail in the present situation). Similar estimates have been used in the context of Variational Inequalities of Evolution by H. Brézis [6]; see Section 1.6.5.

The existence of a solution assumes that the initial density $\rho^{o}(x)$ is bounded and satisfies

$$(*) \qquad \rho^{o}(x) \geq \alpha > 0 \quad .$$

(1) An aspect not considered here.

ISBN 0-12-195250-9

An interesting question arises when $\alpha \to 0$. This problem is solved (in Section 1.6.2) for a penalized model associated to the initial set of equations and using the method of compensated compactness due to Murat and Tartar; a trivial particular case of this useful idea is as follows: if u_α, v_α are sequences of functions weakly convergent in $L^2(\Omega)$, $\Omega \subset \mathbb{R}^2$ and if we know that $\frac{\partial u_\alpha}{\partial x_1} \to \frac{\partial u}{\partial x_1}$ in $L^2(\Omega)$ weakly and that $\frac{\partial v_\alpha}{\partial x_2} \to \frac{\partial v}{\partial x_1}$ in $L^2(\Omega)$ weakly, then $u_\alpha v_\alpha \to uv$ in the sense of distributions in Ω.

In Section 2 we consider a flow of a Newtonian fluid (1) in a domain Ω_ε which consists of a domain Ω with a large number of "small" obstacles (of size ε) arranged in a periodic manner ; we obtain (in a formal manner) an expansion of the velocity u_ε and of the pressure p_ε in terms of ε ; we follow the method of multi-scales as presented in a systematic manner in the book A. Bensoussan, J.L. Lions and G. Papanicolaou [3]. The first term of the expansion obtained in this manner coincides with the first term obtained by H.I. Ene and E. Sanchez-Palencia [10]; the next terms seem to be obtained here for the first time.

The plan is as follows :

---

(1) Similar questions for non newtonian fluids will be considered in latter papers.

1. ON NON HOMOGENEOUS INCOMPRESSIBLE FLOWS.

1.1. Setting of the problem.

Let Ω be a bounded open set of $\mathbb{R}^3$ (1), with boundary Γ (not necessarily smooth). In the cylinder $Q = \Omega \times]0,T[$, $T < \infty$, we consider the system of equations :

$$\rho(\frac{\partial u}{\partial t} + (u.\nabla)u) - \mu\Delta u = \rho f - \nabla p, \tag{1.1}$$

$$\operatorname{div} u = 0 \tag{1.2}$$

$$\frac{\partial \rho}{\partial t} + (u.\nabla)\rho = 0 \tag{1.3}$$

where $u = \{u_i \ ; \ 1 \le i \le 3\}$, ρ and p denote the velocity, the density and the pressure. We have $\mu > 0$; $(u.\nabla)u = u_j \frac{\partial u}{\partial x_j}$ (with the summation convention).

The initial conditions are

$$u(x,0) = u^0(x), \qquad x \in \Omega, \tag{1.4}$$

$$\rho(x,0) = \rho^0(x) \tag{1.5}$$

and the boundary conditions are

$$u = 0 \text{ on } \Sigma, \qquad \Sigma = \Gamma \times]0,T[. \tag{1.6}$$

We are going to show global existence of a weak solution to (1.1) ... (1.6) under standard hypothesis on f and on u^0 and under the hypothesis

$$0 < \alpha \le \rho^0(x) \le \beta < \infty, \tag{1.7}$$

the essential point here being that $\alpha > 0$. We shall study in Section 1.6.2. below the case when α can be zero, on a model modified by penalizations.

1.2. Weak solutions.

By multiplying (1.3) by u and adding to (1.1) we can write

$$\frac{\partial}{\partial t}(\rho u) + \frac{\partial}{\partial x_j}(u_j \rho u) - \mu\Delta u = \rho f - \nabla p, \tag{1.8}$$

---

(1) The results to follow are valid if $\Omega \subset \mathbb{R}^2$, together with some supplementary properties. Cf. Remark 1.3 below.

and (1.3) can be written as well

$$(1.9) \qquad \frac{\partial \rho}{\partial t} + \operatorname{div}(\rho u) = 0.$$

Under this "conservative" form, it is immediate to define <u>weak solutions</u> of the problem.

Let us introduce

$$\mathcal{V} = \{v \mid v \in \mathcal{D}(\Omega)^3 \text{ (}^1\text{)}, \ \operatorname{div} v = 0\},$$

$$V = \text{closure of } \mathcal{V} \text{ in } H^1(\Omega)^3, \text{ (}^2\text{)}$$

$$H = \text{closure of } \mathcal{V} \text{ in } L^2(\Omega)^3.$$

<u>The "weak problem" is now</u> : find u and ρ such that

$$(1.10) \qquad u \in L^2(0,T;V), \quad \rho \in L^\infty(Q),$$

$$(1.11) \qquad \int_0^T (\rho u, -\frac{\partial \varphi}{\partial t})\, dt - \iint_Q u_j \rho u \frac{\partial \varphi}{\partial x_j}\, dx\, dt + \int_0^T a(u,\varphi)\, dt = \\ = \int_0^T (\rho f, \varphi)\, dt + (\rho^0 u^0, \varphi(x,0))$$

$\forall\, \varphi$ smooth vector such that $\operatorname{div}_x \varphi = 0$ and $\varphi(x,T) = 0$, and ρ satisfying (1.9) and (1.5). ■

In (1.11), $(\varphi,\psi) = \int_\Omega \varphi_j \psi_j\, dx$,

$$a(u,\varphi) = \mu \int_\Omega \frac{\partial u_i}{\partial x_j} \frac{\partial \varphi_i}{\partial x_j}\, dx.$$

In (1.11) the initial condition is expressed in a weak form ; we return to that in Remark 1.4 below.

It follows from (1.9) and (1.10) that

$$(1.12) \qquad \frac{\partial \rho}{\partial t} \in L^2(0,T;H^{-1}(\Omega)) \text{ (}^3\text{)}$$

so that ρ is (a.e. equal to) a continuous function from $[0,T] \to H^{-\frac{1}{2}}(\Omega)$ (^{4}), and (1.5) makes sense.

(^{1}) $\mathcal{D}(\Omega)$ = space of $\mathcal{C}^\infty$ functions with compact support. Allfunctions are <u>real valued</u>.

(^{2}) $H^1(\Omega)$ denotes the Sobolev space of functions $\varphi \in L^2(\Omega)$ such that $\frac{\partial \varphi}{\partial x_i} \in L^2(\Omega)\ \forall i$.

(^{3}) $H^{-1}(\Omega)$ = dual space of $H_0^1(\Omega)$, $H_0^1(\Omega)$ = closure of $\mathcal{D}(\Omega)$ in $H^1(\Omega)$.

(^{4}) We refer, for instance, to J.L.Lions, E.Magenes [24] for spaces $H^s(\Omega)$, $s \in \mathbb{R}$.

Remark 1.1

If $\rho^o(x)$ = constant, say $\rho^o=1$, then $\rho = 1$ satisfies (1.9) and (1.11) reduces to the classical formulation of weak solutions of Navier Stokes equations (cf. J. Leray [16][17] [18], E. Hopf [11], O.A. Ladyzenskaya [15]).

1.3. Existence theorem.

THEOREM 1.1. (Kajikov [12], Antonzev and Kajikov [1]). We assume that $\Omega \subset \mathbb{R}^3$ and that

$$f \in L^2(0,T;H) \quad , \tag{1.13}$$

$$u^o \in H \tag{1.14}$$

and that ρ^o satisfies (1.7). Then there exists a weak solution (in the sense of Section 1.2) which satisfies moreover

$$u \in L^\infty(0,T;H). \tag{1.15}$$

Remark 1.2.

Uniqueness is an open question. Let us remark that the question of Uniqueness when $\Omega \subset \mathbb{R}^3$ is still open for the classical Navier Stokes equations. ∎

Remark 1.3. The case $\Omega \subset \mathbb{R}^2$.

If $\Omega \subset \mathbb{R}^2$ one can prove (cf. Antonzev and Kajikov, loc.cit) the existence of stronger solutions :

THEOREM 1.2. The hypothesis are those of Theorem 1.1. ; moreover $\Omega \subset \mathbb{R}^2$ and

$$u^o \in V. \tag{1.16}$$

Then there exists a solution which satisfies

$$\frac{\partial^2 u_i}{\partial x_j \partial x_k} \; , \; \frac{\partial u_i}{\partial t} \in L^2(Q) \quad \forall i,j,k. \tag{1.17}$$

(It follows easily from these properties that $\nabla p \in (L^2(Q))^2$). ∎

The problem of uniqueness.

Uniqueness is an open question even in the class of "strong" solutions of Theorem 1.2, if $\Omega \subset \mathbb{R}^2$. But the existence and uniqueness has been proven by Kajikov [13] in the class of solutions $u \in \mathcal{C}^{2+\beta, 1+\beta/2}$, $\rho \in \mathcal{C}^1(\bar{\Omega} \times]0,T[)$ (under natural smoothness and compatibility hypothesis).Cf. also [33]. In the case ρ = constant, uniqueness of weak solutions is known (cf. J.L. Lions and G. Prodi [25].

Remark 1.4.

One can define V^σ, $\forall \sigma \in \mathbb{R}$, as the closure of $\mathcal{V}$ in $(H^\sigma(\Omega))^3$. Let us remark that (by interpolation estimates)

$$L^2(0,T;V) \cap L^\infty(0,T;H) \subset L^{2/\theta}(0,T;V^\theta). \tag{1.18}$$

By the "fractional Sobolev" imbedding theorem

$$V^\theta \subset (L^{q(\theta)}(\Omega))^3 \ , \quad \frac{1}{q(\theta)} = \frac{1}{2} - \frac{\theta}{3}. \tag{1.19}$$

Taking $\theta = \frac{1}{2}$, we have in particular that

$$u_i \in L^4(0,T;\ L^3(\Omega)) \qquad \forall i\ . \tag{1.20}$$

It follows then from (1.11) that

$$\frac{\partial}{\partial t}(\rho u) \in L^2(0,T\ ;\ V^{-3/2}) \tag{1.21}$$

so that ρu is, in particular, a.e. equal to a continuous function from $[0,T] \to V^{-3/4}$ so that (1.11) implies

$$(\rho u)_{t=0} = \rho^0 u^0. \tag{1.22}$$

1.4. Proof of existence (I). Semi-Galerkin approximation.

We consider an "approximation" of (1.1)(1.2)(1.3) which is of the Galerkin's type in u and where in (1.3) we replace u by its "approximation" (hence the terminology of"Semi-Galerkin"). More precisely, we introduce :

$$\left\{ \begin{array}{l} V_m \subset V\ , \quad V_m \text{ finite dimensional, "}V_m \to V\text{" as} \\ m \to \infty \text{ i.e. } \forall v \in V \text{ there exists } v_m \in V_m \text{ such} \\ \text{that } v_m \to v \text{ in } V \text{ as } m \to \infty. \end{array} \right. \tag{1.23}$$

We are looking for u_m, ρ_m such that

(1.24) $$u_m(t) \in V_m \quad \forall t \ [1]$$

(1.25) $$\begin{cases} (\rho_m \frac{\partial u_m}{\partial t}, v) + \int_\Omega \rho_m \, u_{jm} (\frac{\partial u_{im}}{\partial x_j}) \, v_i dx + a(u_m, v) = \\ = (\rho_m f, v) \quad \forall \, v \in V_m \, , [2] \end{cases}$$

(1.26) $$\frac{\partial \rho_m}{\partial t} + u_{jm} \frac{\partial \rho_m}{\partial x_j} = 0 \, ,$$

(1.27) $$u_m(o) = u_m^o \, , \; u_m^o \in V_m, \quad u_m^o \to u^o \text{ in } H \text{ as } m \to \infty \, ,$$

(1.28) $$\rho_m(o) = \rho^o.$$ ∎

It immediately follows from (1.26) that ρ_m is constant along the trajectories of the particles, so that

(1.29) $$\alpha \le \rho_m(x,t) \le \beta$$

whenever ρ_m exists.

If we take $v = u_m$ in (1.25) and if we multiply (1.26) by $|u_m|^2$, we obtain after adding up :

(1.30) $$\frac{d}{dt} \int_\Omega \rho_m \frac{u_m^2}{2} dx + \int_\Omega \frac{\partial}{\partial x_j} (u_{jm} \rho_m \tfrac{1}{2} u_m^2) dx + a(u_m, u_m) = (\rho_m f, u_m) \, ,$$

or, since $u_m = 0$ on Γ :

(1.31) $$\frac{d}{dt} \int_\Omega \rho_m \frac{u_m^2}{2} dx + a(u_m, u_m) = (\rho_m f, u_m) \, .$$

One can show that it follows from (1.31) that ρ_m and u_m are globally defined in t and that, as $m \to \infty$,

(1.32) u_m is bounded in $L^2(O,T;V) \cap L^\infty(O,T;H)$ [3]. ∎

---

[1] $u_m(t)$ denotes the function $x \quad u_m(x,t)$.

[2] We assume that all functions of V_m are smooth so that all integrals make sense. Such spaces V_m exist.

[3] We obtain that $\int_\Omega \rho_m \, u_m^2 \, dx$ is bounded in t and <u>since</u> $\rho_m \ge \alpha > 0$, u_m is bounded in $L^\infty(O,T;H)$.

Estimates on t-derivatives, or, at least, some <u>fractional</u> t-<u>derivatives</u> are necessary in order to be able to pass to the limit in m, by <u>compactness arguments</u> (cf. for instance, J.L. Lions [20]).

1.5. <u>Proof of existence</u> (II). <u>Estimates on fractional t-derivatives</u>.

It immediately follows from (1.26) that

(1.33) $$\frac{\partial\rho_m}{\partial t} \text{ is bounded in } L^2(0,T;H^{-1}(\Omega)).$$

We multiply (1.26) by $u_m.v$ and we integrate over Ω and we add the result to (1.25) ; we obtain:

$$(\frac{\partial}{\partial t}(\rho_m u_m),v)+\int_\Omega\left[\rho_m u_{jm}\frac{\partial u_{im}}{\partial x_j}v_i+\frac{\partial\rho_m}{\partial x_j}u_{jm}u_{im}v_i\right]dx+a(u_m,v) =$$

$$= (\rho_m f,v)$$

hence

(1.34) $$(\frac{\partial}{\partial t}(\rho_m u_m),v) = (F_m(t),v) \quad \forall\, v\in V_m$$

where

(1.35) $$(F_m,v)=(\rho_m f,v)-a(u_m,v)+\int_\Omega \rho_m u_{im}u_{jm}\frac{\partial v_i}{\partial x_j}dx.$$

One easily verifies that

(1.36) $$|(F_m(t),v)|\leq c[k(t)+\|u_m(t)\|^2] \quad \|v\|$$

where here and in what follows the c's denote various constants and where $k\in L^2(0,T)$ and where

$$\| \; \| = \text{norm in } V.$$

One can deduce from (1.34) that [1]

(1.37) $$\begin{cases} \text{there exists a constant } c \text{ such that} \\ \int_0^{T-\delta}|u_m(t+\delta)-u_m(t)|^2 dt\leq c\delta^{\frac{1}{2}} \quad \forall\delta>0. \end{cases}$$

---

[1] We denote by $|\;|$ the norm in H.

Sketch of proof.

We integrate (1.34) in t from t to $t+\delta$, and we take in the result $v=u_m(t+\delta)-u_m(t)$; it comes

$$(1.38)\quad \begin{cases} ((\rho_m u_m)(t+\delta)-(\rho_m u_m)(t),\ u_m(t+\delta)-u_m(t)) = \\ =\left(\int_t^{t+\delta} F_m(s)ds,\ u_m(t+\delta)-u_m(t)\right). \end{cases}$$

Let us set

$$X_m=(\rho_m(t+\delta)(u_m(t+\delta)-u_m(t)),\ u_m(t+\delta)-u_m(t)),$$

$$Y_m=((\rho_m(t+\delta)-\rho_m(t))u_m(t),\ u_m(t+\delta)-u_m(t)).$$

There (1.38) can be written

$$(1.39)\quad X_m = -Y_m+\left(\int_t^{t+\delta} F_m(s)ds,\ u_m(t+\delta)-u_m(t)\right).$$

But it follows from (1.26) that

$$((\rho_m(t+\delta)-\rho_m(t))\ u_m(t),v) = \left(\int_t^{t+\delta} u_{jm}\rho_m ds,\ \frac{\partial}{\partial x_j}(u_m v)\right)$$

so that

$$|(\rho_m(t+\delta)-\rho_m(t))\ u_m(t),v)|\leq c\int_t^{t+\delta}\|u_m(s)\|_{(L^4(\Omega))^3} ds\ \times$$

$$\times\left[\|u_m(t)\|_{(L^4(\Omega))^3}\|v\| + \|u_m(t)\|\|v\|_{(L^4(\Omega))^3}\right];$$

since, in particular, $V\subset(L^4(\Omega))^3$, it follows that

$$|Y_m|\leq c\left(\int_t^{t+\delta}\|u_m(s)\|ds\right)\|u_m(t)\|\left[\|u_m(t+\delta)\|+\|u_m(t)\|\right]$$

$$\leq c\ \delta^{\frac{1}{2}}\left(\int_t^{t+\delta}\|u_m(s)\|^2ds\right)^{\frac{1}{2}}\left[\|u_m(t)\|^2+\|u_m(t+\delta)\|^2\right]$$

hence it follows that

$$(1.40)\quad \int_o^{T-\delta}|Y_m|dt\leq c\ \delta^{\frac{1}{2}}\ .$$

Using (1.36) one verifies that

$$\int_o^{T-\delta}\left|\left(\int_t^{t+\delta}F_m ds,\ u_m(t+\delta)-u_m(t)\right)\right|dt\leq c\ \delta^{\frac{1}{2}}$$

so that (1.39) implies

$$(1.41)\quad \int_o^{T-\delta}|X_m|dt\leq C\,\delta^{\frac{1}{2}}\ .$$

But, since $\rho_m \geq \alpha > 0$, (1.41) implies

$$\int_0^{T-\delta} |u_m(t+\delta)-u_m(t)|^2 dt \leq C\,\delta^{\frac{1}{2}} \quad . \tag{1.42}$$

■

Remark 1.5.

Estimates of the type (1.42) have been obtained by H. Brézis [6] for solving variational inequalities of evolution (Cf. Section 1.6.5). ■

These estimates permit to pass to the limit using compactness arguments. By standard arguments, one proves that

$$\left\{\begin{array}{l} u_m \ \text{remains in a compact set of} \\ L^p(0,T;(L^q(\Omega))^3) \text{ if } p\in[2,\infty[,\ q\in[2,6[, \\ \frac{1}{p}+\frac{3}{2q}>\frac{3}{4}\ . \end{array}\right. \tag{1.43}$$

(cf. Antonzev and Kajikov [1], and also a proof in J.L.Lions [21]).

It is then a straightforward matter to verify that one can extract from $\{u_m,\rho_m\}$ a sequence which converges (in weak or weak star topologies) to a weak solution of the problem. ■

1.6. Variants and open questions.

1.6.1. Non-homogeneous Bingham's fluids.

We have studied in G. Duvaut and J.L. Lions [9] flows of homogeneous Bingham's fluids, using Variational Inequalities.

Non-homogeneous Bingham's fluids can be modelled by

$$\rho\left(\frac{\partial u_i}{\partial t}+u_j\frac{\partial u_i}{\partial x_j}\right)-\frac{\partial}{\partial x_j}\sigma^D_{ij}=\rho f_i-\frac{\partial p}{\partial x_i}\ , \tag{1.44}$$

where

$$\left\{\begin{array}{l} \sigma^D_{ij}=\left(2\mu+\dfrac{g}{D_{II}(u)}\right)D_{ij}(u) \text{ if } D_{II}(u)\neq 0\ , \\ \frac{1}{2}\sigma^D_{ij}\,\sigma^D_{ij}\leq g^2 \quad \text{if } D_{II}(u)=0\ , \end{array}\right. \tag{1.45}$$

where

$$D_{II}(u) = \frac{1}{2} D_{ij}(u)\, D_{ij}(u) \ ,$$

$$D_{ij}(v) = \frac{1}{2}\left(\frac{\partial v_i}{\partial x_j} + \frac{\partial v_j}{\partial x_i}\right).$$

The other equations are unchanged, i.e. (1.2)...(1.6).

One can define <u>weak solutions</u> in the following manner. We introduce

$$(1.46) \quad \begin{cases} a(u,v) = 2\mu\,(D_{ij}(u),\, D_{ij}(v)) \ , \\ j(v) = g \displaystyle\int_\Omega (D_{II}(v))^{\frac{1}{2}}\,dx \ . \end{cases}$$

One verifies that if u is a <u>smooth</u> solution of (1.44) (1.45) (together with (1.2) and (1.6)), then

$$(1.47) \quad \begin{cases} (\rho\frac{\partial u}{\partial t}, v-u) + (\rho(u.\nabla)u, v-u) + a(u,v-u) + j(v) - j(u) \geq (\rho f, v-u) \\ \qquad\qquad\qquad\qquad\qquad\qquad\qquad \forall\ v \in \mathcal{V}. \end{cases}$$

We multiply (1.3) by $u(v-u)$ and we add up to (1.47). After some computations, one obtains

$$\int_0^T (\rho u, \frac{\partial v}{\partial t})\,dt - \iint_Q \rho u_i u_j \frac{\partial v_i}{\partial x_j} dxdt + \int_0^T [a(u,v-u) + j(v) - j(u)]\,dt \geq$$

$$\geq \int_0^T (\rho f, v-u)\,dt + \frac{1}{2}(\rho(T)u(T), u(T)) - (\rho(T)u(T), v(T))) -$$

$$- \frac{1}{2}(\rho^o u^o, u^o) + (\rho^o u^o, v(o)) =$$

$$= \int_0^T (\rho f, v-u)\,dt + \frac{1}{2}(\rho(T)(v(T)-u(T)), v(T)-u(T)) - \frac{1}{2}(\rho(T)v(T), v(T)) -$$

$$- \frac{1}{2}(\rho^o u^o, u^o) + (\rho^o u^o, v(o)).$$

We shall say that $\{u,\rho\}$ is a <u>weak solution</u> of the non-homogeneous Bingham's fluid if it satisfies (1.10) and

$$(1.48) \quad \begin{aligned} &\int_0^T (\rho u, \frac{\partial v}{\partial t})\,dt - \iint_Q \rho u_i u_j \frac{\partial v_i}{\partial x_j}\, dxdt + \int_0^T [a(u,v-u) + j(v) - j(u)]\,dt \geq \\ &\geq \int_0^T (\rho f, v-u)\,dt - \frac{1}{2}(\rho(T)v(T), v(T)) - \frac{1}{2}(\rho^o u^o, u^o) + (\rho^o u^o, v(o)) \end{aligned}$$

$\forall v$ which is a smooth function in t with values in $\mathcal{V}$, and if one has (1.2)...(1.6).

By combining the methods of G. Duvaut and J.L.Lions, loc. cit., and the above method, one can show existence of a weak solution satisfying (1.15).

Remark 1.6.

If ρ = constant, the above variational inequality reduces to the one studied in Duvaut-Lions, loc.cit. ■

Remark 1.7.

If $g = 0$, the above problem reduces to the one studied before in this paper. ■

Remark 1.8.

The uniqueness is an open problem in all dimensions (it is proven in G. Duvaut-J.L.Lions, loc.cit. if $\Omega \subset \mathbb{R}^2$ and if ρ = = constant). It would be very interesting to see if the results of Kajikov [13] extend to this situation, if $\Omega \subset \mathbb{R}^2$, $\rho \neq$ constant ■

Remark 1.9.

One can study by similar methods the "non-homogeneous" problems associated with the models considered in D.Cioranescu [7] and in J.P.Dias [8]. ■

1.6.2. The problem when $\rho^0 \geq 0$.

The above analysis supposed that $\rho^0(x) \geq \alpha > 0$. Let us now suppose that

$$0 \leq \rho^0(x) \leq \beta < \infty \quad . \tag{1.49}$$

In that case problem (1.1)...(1.6) is open.

But let us introduce the following penalized model :

given $\varepsilon > 0$ and "small", we consider the equations

$$\begin{cases} \rho \frac{\partial u}{\partial t} + (u.\nabla)u) + \rho u \operatorname{div} u + \frac{1}{2}\nabla(\rho|u|^2) - \frac{1}{\varepsilon}\nabla \operatorname{div} u - \mu\Delta u = \rho f, \\ \text{with } (1.3)\dots(1.6). \end{cases} \tag{1.50}$$

If $\operatorname{div} u = 0$, (1.50) reduces to (1.1).

Weak solutions can be defined as in Section 1.2.One can prove :

THEOREM 1.3. Let $\varepsilon > 0$ be fixed. The hypothesis are those of Theorem 1.1., but with (1.49) instead of (1.7). Then problem (1.50)(1.3)...(1.6) admits a weak solution such that u satisfies (1.15).

Sketch of the proof.

1). One replaces first ρ^0 by $\rho^0_\alpha = \rho^0 + \alpha$, $\alpha > 0$. By methods entirely similar to those of Sections 1.4, 1.5,, one proves the existence of u_α, ρ_α solution of the analogous problem with ρ^0 replaced by ρ^0_α. Therefore

$$(1.51)\quad \frac{\partial}{\partial t}(\rho_\alpha u_\alpha) + \frac{\partial}{\partial x_j}(\rho_\alpha u_{\alpha j} u_\alpha) + \frac{1}{2}\nabla(\rho_\alpha |u_\alpha|^2) - \frac{1}{\varepsilon}\nabla(\mathrm{div}\, u_\alpha) - \mu \Delta u_\alpha = \rho_\alpha f,$$

$$(1.52)\quad \frac{\partial \rho_\alpha}{\partial t} + (u_\alpha . \nabla)\rho_\alpha = 0 ;$$

moreover

$$(1.53)\quad \alpha \le \rho_\alpha(x) \le \beta + \alpha \ , \ \sqrt{\rho_\alpha}\ u_{\alpha j} \text{ is bounded in } L^2(Q),$$

$$(1.54)\quad u_\alpha \text{ remains in a bounded set of } L^2(0,T;V) \text{ as } \alpha \to 0.$$

2). Supplementary estimates.

The equation (1.52) can be written

$$(1.55)\quad \frac{\partial \rho_\alpha}{\partial t} + \frac{\partial}{\partial x_j}(u_{\alpha j}\rho_\alpha) - \rho_\alpha \ \mathrm{div}\, u_\alpha = 0$$

which implies, using (1.53)(1.54), that

$$(1.56)\quad \frac{\partial \rho_\alpha}{\partial t} \text{ is bounded in } L^2(0,T;H^{-1}(\Omega)).$$

It follows from (1.51) that

$$(1.57)\quad \frac{\partial}{\partial t}(\rho_\alpha u_\alpha) \text{ is bounded in } L^2(0,T;(H^{-3/2}(\Omega))^3).$$

3). One can then let $\alpha \to 0$ and pass to the limit using an argument of compensated compactness [1].

It follows from (1.54) and (1.53) that $\rho_\alpha u_\alpha$ is bounded in $L^2(0,T;(L^6(\Omega))^3)$; we can therefore extract a subsequence, still denoted by u_α, ρ_α, such that

---

[1] Compensated compactness has been introduced by Murat and Tartar for the study of operators with weak-star convergent coefficients. Cf.F Murat [27], L.Tartar [28]. These ideas can be applied for the solution of non linear evolution problems : a simple example was given in J.L.Lions [22]. For further work along these lines we refer to L. Tartar [29]. Cf. also Ball [2].

(1.58) $u_\alpha \rightarrow u$ in $L^2(0,T;V)$ weakly,

(1.59) $\rho_\alpha \rightarrow \rho$ in $L^\infty(Q)$ weak star,

(1.60) $u_\alpha \rho_\alpha \rightarrow \sigma$ in $L^2(0,T;(L^6(\Omega))^3)$ weakly.

But (1.54)(1.56) and compensated compactness imply that

(1.61) $$\sigma = \rho u.$$

Similarly (1.54) and (1.57) and compensated compactness imply that

(1.62) $(\rho_\alpha u_\alpha) u_{\alpha j} \rightarrow \sigma u_j$ in the sense of distributions in Q,

and since we have already (1.61), we can pass to the limit in (1.51).

It follows from (1.55) that

(1.63) $\frac{\partial}{\partial t} \frac{\partial \rho_\alpha}{\partial x_i}$ is bounded in $L^2(0,T;H^{-2}(\Omega))$

so that (1.54)(1.63) and compensated compactness imply that

(1.64) $u_{\alpha j} \frac{\partial \rho_\alpha}{\partial x_j} \rightarrow u_j \frac{\partial \rho}{\partial x_j}$ in the sense of distributions in Q,

and one can pass to the limit in (1.52). ■

1.6.3. The case $\mu = 0$.

Results on problem (1.1)(1.2)(1.3) when $\mu = 0$ (and (1.6) being replaced by $u.\nu = 0$ on Σ, ν= normal to Γ) are announced in J. Marsden [26], using methods of differential geometry. It would be interesting to find a direct solution (a proof of a local (in time) existence theorem), analogous to the solution given in Temam [30] for the case where ρ= constant. Cf. also [32] and [31].

1.6.4. Other models.

We refer to Kajikov and Smagulov [14] for similar results in models with diffusion on ρ.

1.6.5. Unilateral problems and estimates on fractional t-derivatives.

This Section follows H. Brézis [6]. We present the main idea on a very simplified model ; we consider Hilbert spaces V and H [1], $V \subset H$, V dense in H ; the norms in V and in H are respectively denoted by $\| \ \|$ and by $| \ |$; let $a(u,v)$ be a continuous bilinear form on V such that

[1] We can take V and H as in Section 1.2, but the situation is general.

(1.65) $\quad$ a is <u>symmetric</u>, and $a(v,v) \geq \alpha \|v\|^2, \alpha > 0, \forall v \in V$;
let $v \to j(v)$ be a <u>convex function</u> from $V \to [0,+\infty]$, $j(o) = 0$, <u>lower semi continuous</u>. [Main example : $j=0$ on a closed convex set K of V, $o \in K$, and $j = +\infty$ outside K].

Let V' be the dual of V when H is identified to its dual ; given $f \in L^2(0,T;V')$, it is known <u>that one can find a function</u> u <u>which satisfies</u>

$$u \in L^2(0,T;V) \cap L^\infty(0,T;H) \, , \tag{1.66}$$

and which is a "weak solution" of the Variational Inequality ;

$$\left\{ \begin{array}{l} \int_0^T \left(\frac{\partial v}{\partial t}, v-u\right) + a(u,v-u) + j(v) - j(u) - (f,v-u) \, dt \geq 0 \\ \forall \, v \in L^2(0,T;V), \; \frac{\partial v}{\partial t} \in L^2(0,T;V'), v(o)=0, j(v) \in L^1(0,T). \end{array} \right. \tag{1.67}$$

The new fact, observed by H. Brézis independently from Kajikov's work, is <u>that one can find a solution such that</u>

$$\int_h^T |u(t)-u(t-h)|^2 \, dt \;\leq\; c \, h^{\frac{1}{2}} \, . \tag{1.68}$$

<u>Remark 1.10.</u>

Strong solutions (such that, for instance, $\frac{\partial u}{\partial t} \in L^2(0,T;V')$) do not necessarily exist. Moreover the above result is still valid if we take V and H as in Section 1.2 and if we add in (1.67) the term

$$b(u,u,v) = \int_\Omega (u.\nabla)u.v \, dx,$$

in dimension ≤ 4. ■

<u>Proof of (1.68).</u>

1) - Let us assume that u is a strong solution and let us prove that

$$\int_h^T |u(t)-u(t-h)|^2 dt \leq h^{\frac{1}{2}} \Phi\left(\|u\|_{L^2(0,T;V)} \, , \; \|u\|_{L^\infty(0,T;H)} \right) . \tag{1.69}$$

One obtains then (1.68) by passing to the limit, starting with some smooth approximation of the weak variational inequality.

We easily verify that if u satisfies (1.67) and $\frac{\partial u}{\partial t} \in L^2(0,T;V')$ then

$$(1.70) \begin{cases} \left(\frac{\partial u(t)}{\partial t}, v-u(t)\right) + a(u(t),v-u(t)) + j(v) - j(u(t)) \geq (f(t), v-u(t)), \\ \forall\, v \in V \text{ and for a.e.t.} \end{cases}$$

If we introduce

$$(1.71) \qquad \tilde{j}(v) = j(v) + \tfrac{1}{2}a(v,v)$$

then (1.70) <u>implies</u>

$$(1.72) \quad \left(\frac{\partial u(t)}{\partial t}, v-u(t)\right) + \tilde{j}(v) - \tilde{j}(u(t)) \geq (f(t), v-u(t)) \quad \forall v \in V.$$

[Indeed $a(u,v-u) \leq \frac{1}{2}a(v,v) - \frac{1}{2}a(u,u)$ since $\frac{1}{2}a(v-u,v-u) \geq 0$].

If we take $v=0$ in (1.72) we obtain (since $\tilde{j}(0)=0$) :

$$\int_0^T \tilde{j}(u)\,dt + \int_0^T \left(\frac{\partial u(t)}{\partial t}, u(t)\right) dt \leq \int_0^T (f(t),u(t))\,dt$$

hence

$$(1.73) \qquad \int_0^T \tilde{j}(u)\,dt \leq c\, \|u\|_{L^2(0,T;V)} .$$

2) - <u>The main point of the proof is now the following</u> :

for $h>0$ and $t \geq h$, we define

$$(1.74) \qquad u_h(t) = h^{-1} \int_{t-h}^{t} u(s)\,ds$$

and we take $v=u_h(t)$ in (1.72) and we integrate over (h,T).
We set :

$$X = \int_h^T \left(\frac{\partial u}{\partial t}, u_h - u\right) dt ,$$

$$Y = \int_h^T [\tilde{j}(u_h(t)) - \tilde{j}(u(t))]\,dt.$$

We have

$$(1.75) \qquad X + Y \geq \int_h^T (f, u_h - u)\,dt.$$

We compute X :

$$X = -\frac{1}{2}|u(T)|^2+\frac{1}{2}|u(h)|^2+(u(T),u_h(T))-(u(h),u_h(h)) - \int_h^T \left(u,\frac{\partial u_h}{\partial t}\right) dt .$$

Since $|u(h)|$, $|u(T)|$, $|u_h(T)|$, $|u_h(h)|$ are all $\leq |u|_{L^\infty(O,T;H)}$, we obtain :

$$X \leq \frac{5}{2}|u|^2_{L^\infty(O,T;H)} - \frac{1}{h}\int_h^T (u(t),u(t)-u(t-h))dt.$$

But $(u(t),u(t)-u(t-h)) = \frac{1}{2}\left[|u(t)|^2-|u(t-h)|^2+|u(t)-u(t-h)|^2\right]$

so that

$$-\int_h^T (u(t),u(t)-u(t-h))dt = -\frac{1}{2}\int_{T-h}^T |u(t)|^2dt+\frac{1}{2}\int_o^h |u(t)|^2dt - \frac{1}{2}\int_h^T |u(t)-u(t-h)|^2dt,$$

and finally

$$X\leq 3|u|^2_{L^\infty(O,T;H)} - \frac{1}{2h}\int_h^T |u(t)-u(t-h)|^2dt. \tag{1.76}$$

We now estimate Y ; since $\tilde{j}$ is convex, it follows from Jensen's inequality that :

$$\tilde{j}(u_h(t))\leq \frac{1}{h}\int_{t-h}^t \tilde{j}(u(s))ds$$

so that

$$Y\leq \int_h^T \frac{dt}{h}\int_{t-h}^t \tilde{j}(u(s))ds - \int_h^T \tilde{j}(u(t))dt =$$

$$= -\int_{T-h}^T \tilde{j}(u(s))ds + \int_{T-h}^T \left(\frac{T-s}{h}\right)\tilde{j}(u(s))ds + \int_o^h \frac{s}{h}\tilde{j}(u(s))ds \leq$$

$$\leq \int_o^h \tilde{j}(u(s))ds \leq \|\tilde{j}(u)\|_{L^1(O,T)}$$

so that by (1.73)

$$Y\leq c\,\|u\|_{L^2(O,T;V)} . \tag{1.77}$$

It follows from (1.75)(1.76)(1.77) that

$$(1.78)\qquad \frac{1}{2h}\int_h^T |u(t)-u(t-h)|^2 dt \leq 3|u|^2_{L^\infty(O,T;H)} + c\|u\|_{L^2(O,T;V)} + Z ,$$

$$Z = \left|\int_h^T (f,u_h-u)\,dt\right| .$$

But

$$Z \leq \|f\|_{L^2(O,T;V')}\left[\|u\|_{L^2(O,T;V)} + \|u_h\|_{L^2(h,T;V)}\right];$$

$$\|u_h(t)\| \leq \frac{1}{h}\int_{t-h}^t \|u(s)\|\,ds \leq \frac{1}{h^{\frac{1}{2}}}\left(\int_{t-h}^t \|u(s)\|^2 ds\right)^{\frac{1}{2}} \leq$$

$$\leq \frac{1}{h^{\frac{1}{2}}}\|u\|_{L^2(O,T;V)}$$

so that, in particular, $\|u_h\|_{L^2(h,T;V)} \leq \frac{c}{h^{\frac{1}{2}}}\|u\|_{L^2(O,T;V)}$.

Therefore $Z \leq c(1+\frac{1}{h^{\frac{1}{2}}})\|u\|_{L^2(O,T;V)}$; using this estimate in (1.78) gives the result.

2. ASYMPTOTIC EXPANSIONS OF SOLUTIONS IN FLOWS IN MEDIA WITH PERIODIC OBSTACLES.

2.1. Setting of the problem.

In $\mathbb{R}^n$ we consider the "basic cell" $Y = \prod_{j=1}^n]o,y_j^o[$; let $\mathcal{O}$ be an open set such that $\mathcal{O}\subset Y$ and let S be the boundary of $\mathcal{O}$. We cover $\mathbb{R}^n$ by all the translations of Y by $m_j y_j^o$ in direction y_j, $1\leq j\leq n$, $m_j \in \mathbb{Z}$; let τY be this covering and let $\tau\mathcal{O}$ be the corresponding translations of $\mathcal{O}$. We then define by homothety $\varepsilon\tau Y$ and $\varepsilon\tau\mathcal{O}$.

Let now Ω be a bounded open set of $\mathbb{R}^n$, with boundary Γ. We define

$$\mathcal{O}_\varepsilon = (\varepsilon\tau\mathcal{O})\cap\Omega,$$
$$\Omega_\varepsilon = \Omega\cap\complement\mathcal{O}_\varepsilon ,$$
$$S = \partial\mathcal{O}_\varepsilon ,\quad \Gamma_\varepsilon = \partial\Omega_\varepsilon\cap\Gamma.$$

The set $\mathcal{O}_\varepsilon$ is the set of "holes" or of "obstacles" in Ω. Let u_ε, p_ε be the (or a) solution of the Navier Stokes in Ω_ε, i.e.

(2.1) $\frac{\partial u_\varepsilon}{\partial t} + (u_\varepsilon \cdot \nabla) u_\varepsilon - \mu \Delta u_\varepsilon = f - \nabla p_\varepsilon$ in Ω_ε,

(2.2) $\operatorname{div} u_\varepsilon = 0$ in Ω_ε,

(2.3) $u_\varepsilon = 0$ on $S_\varepsilon \cup \Gamma_\varepsilon$,

(2.4) $u_\varepsilon(x,o) = u^o(x)$ in Ω_ε;

in (2.1) (resp.2.4) f (resp. u^o) is assumed to be given in $\Omega \times]o,T[$ (resp. in Ω).

We want to study the behaviour of u_ε and of p_ε when $\varepsilon \to 0$. ∎

We consider the stationary problem (1) and we confine ourselves to the linear problem (2), i.e.

(2.5) $-\mu \Delta u_\varepsilon = f - \nabla p_\varepsilon$,

(2.6) $\operatorname{div} u_\varepsilon = 0$ in Ω_ε,

(2.7) $u_\varepsilon = 0$ on $S_\varepsilon \cup \Gamma_\varepsilon$.

We normalize p_ε by, say,

(2.8) $\int_{\Omega_\varepsilon} p_\varepsilon(x)\, dx = 0$

and we want to construct an asymptotic expansion of $u_\varepsilon, p_\varepsilon$ in ε.

2.2. Formal structure of the expansion.

We look for u_ε and p_ε in the form (3)

(2.9) $u_\varepsilon(x) = \sum_{j=0}^{\infty} \varepsilon^j u_j(x, x/\varepsilon)$,

(2.10) $p_\varepsilon(x) \quad \sum_{j=0}^{\infty} \varepsilon^j p_j(x, x/\varepsilon)$

where

(2.11) $u_j(x,y)$, $p_j(x,y)$ are vectors and functions which are defined for $x \in \Omega_\varepsilon$, $y \in \mathbb{R}^n$ and are Y-periodic in y (4).

---

(1) Cf. A. Bensoussan, J.L. Lions and G. Papanicolaou [5] for the case of evolution.

(2) Cf. Remark 2.1 below for the non linear stationary problem.

(3) Cf. the book A.Bensoussan, J.L. Lions and G. Papanicolaou (loc. cit) and [4] [5].

(4) I.e. they admit period y_j^o in direction y_j, $1 \leq j \leq u$.

We shall impose the <u>boundary conditions</u>

(2.12) $u_j(x,y) = 0$ if $y \in S_\varepsilon$, $\forall x \in \Omega_\varepsilon$;

these conditions imply $u_\varepsilon = 0$ on S_ε.

We shall return in Section 2.4 on the conditions to impose on $u_j(x,y)$ for $x \in \Gamma_\varepsilon$.

2.3. <u>Analytic computations</u>.

To simplify the notations, let us set

(2.13) $A = -\mu\,\Delta$.

If we observe that $\frac{\partial}{\partial x_j}\Phi(x,x/\varepsilon) = \frac{\partial\Phi}{\partial x_j}(x,y) + \frac{1}{\varepsilon}\frac{\partial}{\partial y_j}\Phi(x,y)$ (where we replace y by x/ε), we see that

$$A = \varepsilon^{-2}A_1 + \varepsilon^{-1}A_2 + \varepsilon^{0}A_3,$$

$$A_1 = -\mu\,\Delta_y,$$

$$A_2 = -2\mu\,\frac{\partial^2}{\partial x_j\,\partial y_j}, \quad A_3 = -\mu\,\Delta_x.$$

We observe that

$$\nabla = \varepsilon^{-1}\nabla_y + \nabla_x, \quad \mathrm{div} = \varepsilon^{-1}\mathrm{div}_y + \mathrm{div}_x.$$

With these remarks, and using (2.9)(2.10) in (2.5)(2.6), we obtain :

(2.14) $$A_1\,u_o = 0,$$

(2.15) $$A_1\,u_1 + A_2\,u_o = -\nabla_y p_o, \quad \mathrm{div}_y u_1 + \mathrm{div}_x u_o = 0,$$

(2.16) $$\begin{cases} A_1u_2 + A_2u_1 + A_3u_o = f - \nabla_y p_1 - \nabla_x p_o, \\ \mathrm{div}_y\,u_2 + \mathrm{div}_x\,u_1 = 0, \end{cases}$$

(2.17) $$\begin{cases} A_1u_3 + A_2u_2 + A_3u_1 = -\nabla_y p_2 - \nabla_x p_1, \\ \mathrm{div}_y u_3 + \mathrm{div}_x u_2 = 0 \end{cases}$$

etc.

We see immediately that (2.14) together with (2.12) for j=0 and the Y-periodicity of u_o imply

$$u_o \equiv 0.$$

Then (2.15) reduces to

$$A_1 u_1 = -\nabla_y p_o, \quad div_y u_1 = 0 , \tag{2.18}$$

where u_1 satisfies (2.12) (j=1) and is Y-periodic, and where p_o is Y-periodic. Then

$u_1 \equiv 0$ and $p_o = p_o(x)$ (does not depend on y).

Then (2.16) reduces to

$$\begin{cases} A_1 u_2 = f - \nabla_y p_1 - \nabla_x p_o, \\ div_y u_2 = 0. \end{cases} \tag{2.19}$$

We can compute u_2, p_1 in terms of p_o (still undefined) as follows. In the basic cell Y we consider the equation

$$\begin{cases} -\mu \Delta_y \Phi_\lambda = \lambda - \nabla_y \pi_\lambda \quad \text{in } Y - \mathcal{O} , \\ div_y \Phi_\lambda = 0 \quad \text{in } Y - \mathcal{O}, \\ \Phi_\lambda \text{ and } \pi_\lambda \text{ are Y-periodic}, \\ \Phi_\lambda = 0 \text{ on } \partial\mathcal{O}; \end{cases} \tag{2.20}$$

in (2.20) λ is given in $\mathbb{R}^n$; the problem (2.20) admits a unique solution (if we normalize π_λ by, say, $\int_{Y-\mathcal{O}} \pi_\lambda(y)dy=0$).

This solution depends linearly on λ ; we can write

$$\Phi_\lambda(y) = \Phi(y)\lambda, \quad \Phi(y) = \|\Phi_{ij}(y)\| \in \mathcal{L}(\mathbb{R}^n) , \tag{2.21}$$

$$\pi_\lambda(y) = \omega(y)\lambda, \quad \omega(y) = \{\omega_j(y)\} \in \mathbb{R}^n. \tag{2.22}$$

Then

$$u_2(x,y) = \Phi(y).(f(x) - \nabla_x p_o(x)) , \tag{2.23}$$

$$p_1(x,y) = \omega(y).(f(x) - \nabla_x p_o(x)) + \tilde{p}_1(x). \tag{2.24}$$ ■

In order to obtain equations for p_o, $\tilde{p}_1$, we proceed with the computation. We observe that (2.17) reduces to

$$\begin{aligned} A_1 u_3 &= - A_1 u_2 - \nabla_x p_1 - \nabla_y p_2, \\ div_y u_3 &= - div_x u_2. \end{aligned} \tag{2.25}$$

This problem (in u_3, p_2) admits a solution iff

$$\int_{Y-\mathcal{O}} div_x u_2 \, dy = 0. \tag{2.26}$$

Replacing in (2.26) u_2 by its value (2.23) gives an equation for p_o [1] :

$$(2.27) \qquad - \operatorname{div} \mathcal{M}(\Phi)\nabla_x p_o = - \operatorname{div} \mathcal{M}(\Phi) f \quad \text{in } \Omega$$

where

$$(2.28) \qquad \mathcal{M}(\Phi) = \frac{1}{|Y-\mathcal{O}|} \int_{Y-\mathcal{O}} \Phi(y)\,dy \in \mathcal{L}(\mathbb{R}^n).$$

<u>This is an elliptic equation in</u> Ω <u>with symmetric coefficients</u> ; indeed if $\tilde{\lambda}$ is another element of $\mathbb{R}^n$ and if $\Phi_{\tilde{\lambda}}$, $\pi_{\tilde{\lambda}}$ denotes the corresponding solution, we have, by taking the scalar product of (2.20) with $\Phi_{\tilde{\lambda}}$:

$$(2.29) \qquad |Y-\mathcal{O}| \ \lambda\, \mathcal{M}(\Phi)\tilde{\lambda} = \mu \int_{Y-\mathcal{O}} \frac{\partial \Phi_\lambda}{\partial y_i} \frac{\partial \Phi_{\tilde{\lambda}}}{\partial y_i}\,dy \ ;$$

the symmetry of the matrix $\mathcal{M}(\Phi)$ follows and $\lambda.\mathcal{M}(\Phi)\lambda \geq 0$. If $\lambda.\mathcal{M}(\Phi)\lambda = 0$, then $\frac{\partial \Phi\lambda}{\partial y_i} = 0 \quad \forall i$. Then $\nabla_y \pi_\lambda = \lambda$ and this implies $\lambda = 0$ (π_λ being Y-periodic).

We can then proceed with the computation. We denote by u_3^o, p_2^o the solution of

$$(2.30) \qquad \begin{cases} A_1 u_3^o = A_2 u_2 - \nabla_x \ \omega(y).(f(x) - \nabla_x p_o) - \nabla_y p_2^o \ , \\ \operatorname{div}_y u_3^o = - \operatorname{div}_x u_2 \end{cases}$$

which satisfies $u_3^o = 0$ for $y \in \partial\mathcal{O}$, $x \in \Omega_\varepsilon$ and such that u_3^o and p_2^o are Y-periodic. Then

$$(2.31) \qquad u_3 = u_3^o - \Phi(y) \ \nabla_x \ \tilde{p}_1(x) \ ,$$

$$(2.32) \qquad p_2 = p_2^o - \omega(y) \ \nabla_x \ \tilde{p}_1(x).$$

We obtain an equation for $\tilde{p}_1$ when expressing that the next equation in u_4, p_3 admits a solution, namely that :

$$\int_{Y-\mathcal{O}} \operatorname{div}_x u_3\,dy = 0$$

hence

$$(2.33) \qquad - \operatorname{div} \mathcal{M}(\Phi) \ \nabla_x \ p_1 = - \frac{1}{|Y-\mathcal{O}|} \operatorname{div}_x \int_{Y-\mathcal{O}} u_3^o(x,y)\,dy.$$

---

[1] One deduces from (2.23) that $\frac{1}{|Y-\mathcal{O}|}\int_{Y-\mathcal{O}} u_2(x,y)\,dy = \mathcal{M}(\Phi)(f - \nabla p_o)$ which can be considered as the Darcy's law.

We can then proceed. But of course p_o, $\tilde{p}_1$, ... are not yet defined, by lack of boundary conditions.

2.4. Boundary conditions on p_o, $\tilde{p}_1$, ...

When trying to prove the convergence of the expansion; one sees (cf. A. Bensoussan, J.L. Lions and G. Papanicolaou [5]) that a rather natural condition to impose on the u_j's is

$$\nu(x) \int_{Y-\mathcal{O}} u_j(x,y)\,dy = 0 \text{ on } \Gamma , \tag{2.34}$$

$$\nu(x) = \text{vector normal to } \Gamma \text{ in } x \in \Gamma .$$

It follows that :

$$\nu.\mathfrak{M}(\Phi)\ \nabla_x p_o = \nu.\mathfrak{M}(\Phi) f \quad \text{on } \Gamma , \tag{2.35}$$

$$\nu.\mathfrak{M}(\Phi)\ \nabla_x \tilde{p}_1 = \nu \cdot \frac{1}{|Y-\mathcal{O}|} \int_{Y-\mathcal{O}} u_3^o(x,y)\,dy, \quad x \in \Gamma \tag{2.36}$$

etc.

These are Neumann's boundary conditions.

2.5. Summary of formulas.

To summarize, we have obtained :

$$u_\varepsilon = \varepsilon^2 u_2(x,y) + \varepsilon^3 u_3(x,y) + \dots , \tag{2.37}$$

$$p_\varepsilon = p_o(x) + \varepsilon p_1(x,y) + \dots ; \tag{2.38}$$

the terms of those expansions are computed as follows : one defines $\mathfrak{M}(\Phi)$ by (2.28) and one computes p_o (up to an additive constant) by solving the elliptic equation (2.27) in Ω subject to the Neumann's boundary condition (2.35) on Γ. Then u_2 is given by (2.23). We compute u_3^o , p_2^o by (2.30) and we compute $\tilde{p}_1$ by solving (2.33) subject to (2.36) ; then u_3, p_2 are given by (2.31)(2.32) and we can proceed. ■

We conjecture that

$$u_\varepsilon - \varepsilon^2 u_2(x_1, x/\varepsilon) \to 0 \text{ in } (H^1(\Omega))^3 \tag{2.39}$$

and that, under proper normalizations,

$$p_\varepsilon - p_o - \varepsilon p_1(x, x/\varepsilon) \text{ is } O(\varepsilon^2) \text{ in the } L^2(\Omega) \text{norm.} \tag{2.40}$$

To obtain higher order estimates would probably necessitate boundary corrections in the neighbornood of Γ , but the expansions (2.37)(2.38) are probably valid to any order outside some neighborhood of Γ.

Remark 2.1.

If we consider the non linear stationary problem

$$(2.41)\quad \begin{cases} -\mu\Delta u_\varepsilon + (u_\varepsilon.\nabla)u_\varepsilon = f - \nabla p_\varepsilon , \\ \text{div } u_\varepsilon = 0 , \\ u_\varepsilon = 0 \text{ on } \Gamma_\varepsilon \cup S_\varepsilon . \end{cases}$$

we obtain similar expansions up to the terms u_4 and p_3; only terms u_5 and p_4 are affected by the non linear terms.

REFERENCES

1. S.N. Antonzev and A.V. Kajikov (1973), Mathematical Study of flows of non homogeneous fluids. Novosibirsk Lectures at the University (in Russian).

2. J. Ball (1977), Convexity Conditions and Existence Theorems in non linear Elasticity. A.R.M.A. 63, p. 337-403.

3. A. Bensoussan, J.L. Lions and G. Papanicolaou (1978), Asymptotic Methods in Periodic Structures. North-Holland.

4. ______________________ Asymptotic expansions in perforated media (I). To appear.

5. ______________________ Flows in porous media with periodic structures. To appear.

6. H. Brézis. Personal communication and Unilateral problems in Newtonian fluids.

7. D. Cioranescu (1977), Fluides non-newtoniens and Domaines à cavités, Thesis, Paris.

8. J.P. Dias (1977), Sur les équations couplées bidimensionnelles d'un cristal liquide nématique incompressible. C.R.A.Sc. Paris.

9. G. Duvaut and J.L. Lions (1972), Inequalities in Mechanics and in Physics. Paris, Dunod. (French) ; (1974) Springer.

10. H.I. Ene and E. Sanchez-Palencia (1975), Equations et phénomènes de surface pour l'écoulement dans un modèle de milieu poreux. J. de Mécanique 14, p. 73-108.

11. E. Hopf (1951), Uber die Aufangswertanfgebe für die hydrodynamischen Grundgleichungen. Math. Nachr.4, P.213-231.

12. A.V. Kajikov (1974), Resolution of boundary value problems for non homogeneous viscous fluids. Doklady Akad. Nauk, 216, p. 1008-1010.

13 ________ (1975), In Seminar on Numerical Methods in Mechanics (Sem. N.N.Yanenko), Part II, Novosibirsk, p. 65-76.

14. A.V. Kajikov and Ts. Smagulov, The correctness of a boundary value problem for a model of non homogeneous fluid with diffusion. Doklady Akad. Nauk, 234, p. 330-332, (1977).

15. O.A. Ladyzenskaya (1963), The mathematical theory of viscous incompressible fluids. Gordon Breach. New-York.

16. J. Leray (1933), Etude de diverses équations intégrales non linéaires et de quelques problèmes que pose l'hydrodynamique. J.M.P.A. XII., p. 1 - 82.

17. ________ (1934), Sur le mouvement d'un liquide visqueux emplissant l'espace. Acta Math. 63, p. 193-248.

18. ________ (1934), Essai sur le mouvement plan d'un liquide visqueux que limitent des parois. J.M.P.A.XIII,331-418.

19. T. Lévy and E. Sanchez-Palencia (1975), On boundary conditions for fluid flow in porous media. Int.J. Eng. Sc. 13, p. 923-940.

20. J.L. Lions (1969), Quelques Méthodes de résolution des problèmes aux limites non linéaires. Paris, Dunod Gauthier-Villars.

21. ________ (1977), On some Questions in boundary value problems of Mathematical Physics. Univ. Fed. Rio-de-Janeiro, Instituto de Matematica.

22. ________ (1977), Workshop in Mathematics and Mechanics, Austin, March 1977.

23. ________ (1959), Quelques résultats d'existence dans les équations aux dérivées part. non linéaires. Bull. SMF 87, p. 245-273.

24. J.L. Lions and E. Magenes (1968), Non homogeneous boundary value problems and applications, Vol.I Paris Dunod (Fr.) (1970) Springer (English).

25. J.L. Lions and G. Prodi (1959), Un théorème d'ex stence et unicité dans les éq. de Navier-Stokes en dimension 2. C.R.A.S., Paris, 248, p. 3519-3521.

26. J.E. Marsden (1976), Well posedness of the equations of a non homogeneous perfect fluid. C. in PDE 1(3), 215-230.

27. F. Murat (1977) Sur la compacité compensée. Ann. Scuola Norm. Sup. Pisa, to appear.

28. L. Tartar (1977), Weak convergence in non linear P.D.E. Workshop in Math. and Mechanics, Austin, March 1977.

29. ________ Homogénéisation dans les E.D.P. To appear.

30. R. Temam (1976), Local existence of C^∞ solutions of the Euler eq.in Turbulence of incompressible perfect fluids. In Turbulence & N.S. equations, Springer Lect. Notes in Mathematics, 565, p. 184-194.

31. M.S. Baouendi et C. Goulaouic, (May 1977), Solutions analytiques de l'équation d'Euler d'un fluide compressible. Séminaire Ecole Polytechnique, XXII.

32. O.A. Ladyzenskaya, (1971), On the solution in the small ... Zapiski Nauk Seminar Leningrad, t. 21,5, pp.65-78.

33. O.A. Ladyzenskaya, Personal communication: existence and uniqueness in a Sobolev space $W^{1,p}$ for p large enough has been proved by Ladyzenskaya and Solonnikov.

College de France
Paris, France

Everywhere Defined Wave Operators

Walter A. Strauss

1. INTRODUCTION.

In scattering theory one has a Hilbert space X, a "free" group of unitary operators $U_o(t) = \exp(i\, t\, H_o)$ on X, and a "perturbed" evolution equation. We write the perturbed equation formally as

$$\frac{du}{dt} = i\, H_o u + Pu \tag{1}$$

where P is the perturbation operator. For instance, in classical quantum mechanics, $H_o = -\Delta$ and P is multiplication by a "potential" function $iV(x)$. Here we are concerned with the case when P is nonlinear. One looks for conditions under which there exist free solutions u_+ and u_-, where $u_\pm(t) = \exp(i\, t\, H_o) f_\pm$, and a perturbed solution u such that

$$\|u(t) - u_\pm(t)\| \to 0 \quad \text{as} \quad t \to \pm\infty .$$

The wave operators are defined as $W_\pm : u_\pm(T) \to u(T)$ and the scattering operator as $S : u_-(T) \to u_+(T)$, where T is some convenient reference time.

The existence of the wave operators on a dense subset of X was first proved in realistic nonlinear situations by Segal [9]. More recent expositions may be found in Strauss [12,13] and Reed [7]; see also Strichartz [15]. Years ago Segal suggested the possibility that the wave and scattering operators might be defined on the whole Hilbert space X.

This paper is a step in that direction. We find suf-

ISBN 0-12-195250-9

ficient conditions on X, $U_0(t)$, P under which $W_\pm$ map <u>all</u> of X into X. The main theorem is stated in section 2 and proved in section 3. It is applied in section 4 to the nonlinear Klein-Gordon equation

$$u_{tt} - \Delta u + u + u^3 = 0$$

where $x \in \mathbb{R}^3$ and $X = H^1(\mathbb{R}^3) \oplus L^2(\mathbb{R}^3)$ is the usual Hilbert space of Cauchy data of finite energy. It is applied in section 5 to the nonlinear Schrödinger equation

$$i\, u_t - \Delta u + u^p = 0$$

where $x \in \mathbb{R}^n$ and $p = 1 + 4/n$. Some further remarks are made about this critical power. In section 6 we discuss the existence of the wave operators on dense subsets of X.

Two new developments have made the present results possible. The first is the new decay estimate of Segal [10] for $n = 1$ and Strichartz [16] for general n. They prove that <u>arbitrary</u> finite-energy solutions of the <u>free</u> Klein-Gordon equation belong to L^q of space-time for certain values of q. Similarly, arbitrary solutions of the free Schrödinger equation with initial data in L^2 belong to L^q of space-time for $q = 2 + 4/n$. This is in contrast to the considerably more rapid rate of decay which is valid for only a dense class of solutions. The second new development is a remarkable series of papers of Ginibre and Velo [2,3] on the nonlinear Schrödinger equation. They show how conservation laws can be exploited to enhance weak decay estimates. They are able to treat situations when the perturbation term Pu does not have values in X. Our proof is a modification of some of their techniques.

Segal's original question of whether S is defined on all of X is still open. For the Klein-Gordon case it is conceivable that Morawetz' well-known estimate [5] for perturbed solutions, that $|x|^{-1}|u(x,t)|^4$ is integrable in space-time, may be combined with the present methods to give an affirmative answer. Another natural question is why we do not employ the Lorentz-invariant Hilbert space, which is somewhat bigger than the energy space, for X. The difficulty is that the former space involves the non-local operator $(I - \Delta)^{1/4}$. Thus, for a local perturbation term $F(u(x,t))$, there does not seem to be any conservation law.

In this paper we will use the free group $U_o(t)$ to rewrite (1) as an integral equation. Formally we have

$$u(t) = U_o(t-s)u(s) + \int_s^t U_o(t-\tau)Pu(\tau)d\tau .$$

The advantage is that neither $H_o u(\tau)$ nor $Pu(\tau)$ are required to belong to X. Formally letting $s \to -\infty$, we expect

$$u(t) = u_-(t) + \int_{-\infty}^t U_o(t-\tau)Pu(\tau)d\tau ,$$

our basic equation as regards the wave operator W_-.

A word about notation. If I is an interval of the real line and Z is a Banach space, $C(I,Z)$ denotes the continuous functions from I to Z, and $L^p(I,Z)$ the usual L^p space of functions from I to Z. Various constants are denoted by c.

2. THE MAIN THEOREM.

We make the following hypotheses.

(I) Let X be a Hilbert space with norm $|\ |_2$. Let $U_o(t)$ be a one-parameter group of unitary operators on X.

(II) Let X_1 and X_3 be Banach spaces with norms denoted by $|\ |_1$ and $|\ |_3$, respectively. Let $P : X_3 \to X_1$ be a (nonlinear) operator such that $P0 = 0$ and

$$|Pf - Pg|_1 \le c\,(|f|_3 + |g|_3)^{p-1}|f - g|_3$$

where p and c are constants, $p > 1$.

(III) Let X be continuously embedded as a dense subset of X_3. Let X_3 and X_1 be continuously embedded in some topological linear space.

(IV) For each $f \in X$ assume that the function $t \to U_o(t)f$ belongs to $L^{p+1}(\mathbb{R},X_3)$.

(V) Let $U_o(t)$, restricted to $X \cap X_1$, have a continuous linear extension (still denoted $U_o(t)$) which maps $X_1 \to X_3$ with norm $\le c|t|^{-d}$ for $t \neq 0$. Here c is a constant and $d = 2/(p + 1)$.

(VI) Let G be a continuous non-negative functional on

the space X_3 with the following property. Suppose $I \subset \mathbb{R}$ is some time interval, $s \in I$ and $f \in X$. Whenever $u \in L^{p+1}(I,X_3)$ satisfies the equation

$$u(t) = U_o(t)f + \int_s^t U_o(t-\tau)Pu(\tau)d\tau \tag{2}$$

for a.e. $t \in I$, assume that

(i) $u \in C(I,X)$

(ii) $\frac{1}{2}|u(t)|_2^2 + G(u(t))$ is independent of t

(iii) For all $t,r \in I$,

$$u(t) = U_o(t-r)u(r) + \int_r^t U_o(t-\tau)Pu(\tau)d\tau \quad . \tag{3}$$

THEOREM 1. *If $f_- \in X$ there exists a unique solution of the equation*

$$u(t) = U_o(t)f_- + \int_{-\infty}^t U_o(t-\tau)Pu(\tau)d\tau \tag{4}$$

in some time interval $I = (-\infty,T]$ such that

$$u \in C(I,X) \cap L^{p+1}(I,X_3) \quad ,$$
$$|u(t) - U_o(t)f_-|_2 \to 0 \quad \text{as} \quad t \to -\infty \quad ,$$
$$\frac{1}{2}|u(t)|_2^2 + G(u(t)) = \frac{1}{2}|f_-|_2^2 \quad \text{for} \quad t \in I \quad .$$

LEMMA 1. *Assume (I) - (V). Let I be an interval, $s \in I$ and $u \in L^{p+1}(I,X_3)$. Then the function*

$$v(t) = \int_s^t U_o(t-\tau)Pu(\tau)d\tau$$

also belongs to $L^{p+1}(I,X_3)$. Thus equation (2) makes sense.

Proof. We estimate v as follows.

$$|v(t)|_3 \leq \int_s^t c\ |t-\tau|^{-d}|Pu(\tau)|_1 d\tau$$

where $|Pu(\tau)|_1 \leq c\ |u(\tau)|_3^p$. The assumed relation between d and p is exactly such that the standard singular integral inequality [11] is applicable, using the L^{p+1} norm on $|v(t)|_3$ and the $L^{1+1/p}$ norm on $|u(\tau)|_3^p$.

REMARK 1. If in Hypothesis (VI) we assumed $Pu \in L^1(I,X)$, then (iii) would be trivial. Indeed, we would write (2) with $t = r$,

$$u(r) = U_o(r)f + \int_s^r U_o(r-\tau)Pu(\tau)d\tau$$

and then apply $U_o(t-r)$ to both sides. We could move this operator inside the integral because it converges in the space X. Thus

$$U_o(t-r)u(r) = U_o(t)f + \int_s^r U_o(t-\tau)Pu(\tau)d\tau .$$

If this result is combined with (2), equation (3) follows.

REMARK 2. If we assumed P were a locally Lipschitz map from X to X, then both (i) and (iii) of Hypothesis (VI) would be trivial. That is,

$$|Pf - Pg|_2 \leq b\ |f - g|_2$$

where b depends boundedly on $|f|_2 + |g|_2$. Indeed, in this case we could uniquely solve equation (2) in the space X locally in t by the standard Picard method as in [8]. Then (i) would be valid in a subinterval and, by Remark 1, (iii) would also be valid there. If we assumed (ii) then $|u(t)|_2$ would be bounded, so that $u(t)$ would extend to a solution in $C(\mathbb{R},X)$ of equation (2).

REMARK 3. Hypotheses (II) and (V) will only be used together as in the proof of Lemma 1. Therefore they may be combined into the single hypothesis

$$|U_o(t)(Pf-Pg)|_3 \leq c\ |t|^{-d}(|f|_3 + |g|_3)^{p-1}|f - g|_3$$

for $f,g \in X_3$, $t \neq 0$. The space X_1 has no real significance.

3. PROOF OF THE MAIN THEOREM.

LEMMA 2. *Given* $f_- \in X$, *there exists a finite time* T *such that equation* (4) *has a unique solution* $u \in L^{p+1}(I;X_3)$, *where* I *is the time-interval* $(-\infty,T]$.

Proof. Denote $\mathcal{P}u(t) = \int_{-\infty}^t U_o(t-\tau)Pu(\tau)d\tau$. As in the proof of Lemma 1,

$$|\mathcal{P}u(t)|_3 \le c\int_{-\infty}^{t}(t-\tau)^{-d}|u(\tau)|_3^p d\tau \quad .$$

Denote by $||\ \ ||$ the norm in $L^{p+1}(I,X_3) = B$. Then $||\mathcal{P}u|| \le c\ ||u||^p$ for all $u \in B$. Similarly

$$||\mathcal{P}u-\mathcal{P}v|| \le c(||u||+||v||)^{p-1}||u-v|| \text{ for all } u,v \in B \quad .$$

Let $u_-(t) = U_o(t)f_-$. By Hypothesis (IV), $u_- \in B$. Choose T so small that $c(2||u_-||)^{p-1} < 1/2$. Then $\mathcal{P}$ is a contraction map in the space $\{v \in B : ||v|| \le 2||u_-||\}$. Hence the equation $u = u_- + \mathcal{P}u$ has a unique solution in this space.

LEMMA 3. Let T be chosen as in Lemma 2. Let $s \le T$. Then the equation

$$v_s(t) = U_o(t)f_- + \int_s^t U_o(t-\tau)Pv_s(\tau)d\tau \tag{5}$$

also has a unique solution in $L^{p+1}(I,X_3)$. Furthermore $||v_s - u|| \to 0$ as $s \to -\infty$.

Proof. The proof of the first statement is almost identical with that of Lemma 2. Subtracting equations (4) and (5), we have

$$v_s(t) - u(t) = -\int_{-\infty}^{s} U_o(t-\tau)[Pv_s(\tau) - Pu(\tau)]d\tau \quad .$$

Hence, as before,

$$||v_s - u|| \le c\ (||u||_s + ||v_s||_s)^{p-1}||u - v_s|| \le c||u_-||_s^p$$

where $||w||_s$ denotes the norm in $L^{p+1}((-\infty,s),X_3)$. We have used the inequalities $||u||_s \le 2||u_-||_s$, $||v_s||_s \le 2||u_-||_s$. Since $||u_-||_s \to 0$ as $s \to -\infty$, the proof is complete.

LEMMA 4. Define u as in Lemma 2 and v_s as in Lemma 3. Let $I = (-\infty,T]$. Then $u \in L^\infty(I,X)$ and $v_{s_j} \to u$ weakly* in $L^\infty(I,X)$ for some sequence $s_j \to -\infty$.

Proof. By Hypothesis (VI), $v_s \in C(I,X)$,

$$v_s(t) = U_o(t-r)v_s(r) + \int_r^t U_o(t-\tau)Pv_s(\tau)d\tau \tag{6}$$

and

$$\frac{1}{2}|v_s(t)|_2^2 + G(v_s(t)) = \frac{1}{2}|f_-|_2^2 + G(U_o(s)f_-) \quad . \tag{7}$$

By Hypothesis (IV), there is a sequence $s_j \to -\infty$ such that $|U_o(s_j)f_-|_3 \to 0$. Hence $G(U_o(s_j)f_-) \to 0$. By (7) and the non-negativity of G , $\{v_{s_j}\}$ is a bounded sequence in $L^\infty(I,X)$. Any weak* accumulation point of this sequence in $L^\infty(I,X)$ must be equal to u because of Lemma 2 and $X \subset X_3$. This proves that the sequence itself converges weakly* in $L^\infty(I,X)$ to u .

LEMMA 5. $\frac{1}{2}|u(t)|_2^2 + G(u(t)) \leq \frac{1}{2}|f_-|_2^2$ a.e.

Proof. Let ξ be a non-negative real L^∞ function with compact support in $I = (-\infty,T)$ and integral 1 . Multiplying (7) by $\xi(t)$ and integrating with respect to t, we have

$$\int_{-\infty}^T \{\frac{1}{2}|v_s(t)|_2^2 + G(v_s(t)\}\xi(t)dt = \frac{1}{2}|f_-|_2^2 + G(U_o(s)f_-) \quad .$$

We wish to let $s \to -\infty$. Now $v_s \to u$ strongly in $L^{p+1}(I,X_3)$. So we may pick a sequence $s_j \to -\infty$ such that $U_o(s_j)f_- \to 0$ in X_3 and $|v_{s_j}(t) - u(t)|_3 \to 0$ for a.e. $t \in I$. By Fatou's lemma,

$$\int G(u(t))\xi(t)dt \leq \underline{\lim} \int G(v_{s_j}(t))\xi(t)dt \quad .$$

By Lemma 4,

$$\int |u(t)|_2^2\xi(t)dt \leq \underline{\lim} \int |v_{s_j}(t)|_2^2\xi(t)dt \quad .$$

Also $G(U_o(s_j)f_-) \to 0$. Hence we obtain

$$\int \{\frac{1}{2}|u(t)|_2^2 + G(u(t))\}\xi(t)dt \leq \frac{1}{2}|f_-|_2^2$$

from which the result follows.

LEMMA 6. $u \in C(I,X)$ <u>and</u> u <u>satisfies equation</u> (3).

Proof. We multiply identity (6) by $\xi(t)\eta(r)g$ where ξ and η are bounded real functions of compact support in I and $g \in X_3^*$, the dual space of X_3. Then we integrate over r and t. We let $s = s_j \to -\infty$. Because of Lemma 4, the first term converges to

$$\iint (u(t),g)\xi(t)dt\eta(r)dr \quad .$$

For the same reason the second term converges to

$$\iint (u(r),U_o(r-t)g)\eta(r)dr\xi(t)dt \quad .$$

As for the third term, let

$$y_s(t) = \int_r^t U_o(t-\tau)[Pv_s(\tau) - Pu(\tau)]d\tau \quad ,$$

$$|y_s(t)|_3 \le c \int_{-\infty}^t (t-\tau)^{-d}(|v_s(\tau)|_3 + |u(\tau)|_3)^{p-1}|v_s(\tau) - u(\tau)|_3 d\tau.$$

Because of Lemma 3, $y_s \to 0$ in $L^{p+1}(I,X_3)$ uniformly for $r \in I$. Hence the third term tends to

$$\iint (\int_r^t U_o(t-\tau)Pu(\tau)d\tau,g)\xi(t)dt\eta(r)dr \quad .$$

Since g,ξ and η are arbitrary, the identity (3) follows. It is valid for a.e. $r,t \in I$. By Hypothesis (VI), $u \in C(I,X)$, $|u(t)|_2^2/2 + G(u(t))$ is independent of t and (3) holds for all $t,r \in I$.

LEMMA 7. $\frac{1}{2}|u(t)|_2^2 + G(u(t)) = \frac{1}{2}|f_-|_2^2$.

Proof. Define $w_s(t) = U_o(t-s)u(s)$. By Lemma 6,

$$u(t) = w_s(t) + \int_s^t U_o(t-\tau)Pu(\tau)d\tau \; .$$

Lemma 2 defines u as the solution of (4). Subtracting these two equations,

$$w_s(t) - u_-(t) = \int_{-\infty}^s U_o(t-\tau)Pu(\tau)d\tau \quad .$$

Hence

$$|w_s(t) - u(t)|_3 \le c \int_{-\infty}^s (t-\tau)^{-d}|u(\tau)|_3^p d\tau$$

and

$$\int_{-\infty}^{T} |w_s(t) - u_-(t)|_3^{p+1} dt \leq c\left(\int_{-\infty}^{s} |u(\tau)|_3^{p+1} d\tau\right)^p \to 0$$

as $s \to -\infty$. On the other hand,

$$|w_s(t)|_2 = |w_s(s)|_2 = |u(s)|_2 \leq c$$

for $s \leq T$. Hence $\{w_s\}$ is bounded in $L^\infty(\mathbb{R},X)$. As in the proof of Lemma 4, it follows that $w_s \to u_-$ weakly* in $L^\infty(\mathbb{R},X)$. Hence, with $\xi(t)$ as before,

$$\int |u_-(t)|_2^2 \xi(t) dt \leq \underline{\lim} \int |w_s(t)|_2^2 \xi(t) dt .$$

as $s \to -\infty$. Thus

$$\begin{aligned} |f_-|_2^2 &\leq \underline{\lim}\ |u(s)|_2^2 \leq \overline{\lim} |u(s)|_2^2 \\ &\leq \overline{\lim}\ |u(s)|_2^2 + 2G(u(s)) \leq |f_-|_2^2 . \end{aligned}$$

by Lemma 5. We conclude that $G(u(s)) \to 0$ and $|u(s)|_2 \to |f_-|_2$ as $s \to -\infty$ and Lemma 7 is proved.

The proof of Lemma 7 shows that $w_s \to u_-$ weakly in the Hilbert space $L^2(\mathbb{R},X,\xi)$, which has the norm

$$|v| = \left(\int_{-\infty}^{\infty} |v(t)|_2^2 \xi(t) dt\right)^{1/2} .$$

It also shows that $|w_s| \to |u_-|$ in this norm. Hence $w_s \to u_-$ strongly in this space:

$$|u(s) - u_-(s)|_2^2 = \int |w_s(t) - u_-(t)|_2^2 \xi(t) dt \to 0$$

as $s \to -\infty$. This completes the proof of Theorem 1.

Theorem 2. *Assume* (I), (II), (IV) and (V). *Replace the first sentence in* (III) *by the following*: $X \cap X_3$ *is dense in* X *and in* X_3. *Alter* (VI) *as follows*: G *is a non-negative continuous functional on the space* $X + X_3$. *Replace* (i) *by* $u \in L^\infty(I,X)$. (ii) *and* (iii) *are valid a.e.*

Then Theorem 1 is valid with $C(I,X)$ *replaced by* $L^\infty(I,X)$ *if* t *is excluded from a set of measure zero.*

Proof. The whole proof is unchanged with the following exceptions. At the end of the proof of Lemma 4, we have

$L^\infty(I,X) \subset L^{p+1}(I,X+X_3)$ and $L^{p+1}(I,X_3) \subset L^{p+1}(I,X+X_3)$, so the same conclusion is valid. In Lemma 6, we replace $C(I,X)$ by $L^\infty(I,X)$. In Lemma 7 and in the completion of the proof, a set of measure zero is excluded.

4. THE NONLINEAR KLEIN-GORDON EQUATION.

Consider real-valued solutions of the "perturbed" and "free" equations

$$u_{tt} - \Delta u + m^2 u + \lambda u^3 = 0 \quad , \quad v_{tt} - \Delta v + m^2 v = 0$$

where $x \in \mathbb{R}^3$ and λ and m are positive constants. The energy norm is

$$||u(t)||_e^2 = \int [u_t^2 + |\nabla u|^2 + m^2 u^2] dx \quad .$$

If the energy norm of a solution is finite at one time, it is finite at all times. This is because $||v(t)||_e$ is independent of t for a free solution v and

$$\frac{1}{2}||u(t)||_e^2 + \frac{\lambda}{4}\int u^4 dx$$

is independent of t for a perturbed solution u.

THEOREM 3. *Let u_- be any free solution of finite energy. Then there exists a unique perturbed solution of finite energy with the following properties.*

$$u \in C(\mathbb{R}, W^{1,2}(\mathbb{R}^3)) \quad , \quad u_t \in C(\mathbb{R}, L^2(\mathbb{R}^3)) \quad .$$

$$||u(t) - u_-(t)||_e \to 0 \quad \text{as} \quad t \to -\infty \quad .$$

$$\frac{1}{2}||u(t)||_e^2 + \frac{\lambda}{4}\int u^4 dx = \frac{1}{2}||u_-||_e^2 \quad \text{for all} \quad t \quad .$$

$$\int_{-\infty}^{0}\int_{\mathbb{R}^3} u^4 dx\, dt < \infty \quad .$$

Proof. We denote $L^p = L^p(\mathbb{R}^3)$ and $W^{k,p} = W^{k,p}(\mathbb{R}^3)$ where k denotes the number of derivatives, for the usual Lebesgue and Sobolev spaces. Let $X = W^{1,2} \oplus L^2$, the space of Cauchy data of finite energy, provided with the norm

$$|[f_1,f_2]|_2^2 = \int \{|\nabla f_1|^2 + m^2 f_1^2 + f_2^2\} dx \quad ,$$

where $f = [f_1, f_2]$ denotes an element of X. Let $X_3 = L^4 \oplus W^{-1,4}$. Let $X_1 = \{0\} \oplus L^{4/3}$. Let $p = 3$, $d = 1/2$. Let $P([f_1,f_2]) = [0, \lambda f_1^3]$ be the perturbation operator. Let $G([f_1,f_2]) = \frac{\lambda}{4}\int f_1^4 dx$.

We verify Hypotheses (I)-(VI). (I) is easy. (II) is equivalent to the inequality

$$\left(\int |f_1^3 - g_1^3|^{4/3} dx\right)^3 \le c \left(\int f_1^4 dx + \int g_1^4 dx\right)^2 \int (f_1 - g_1)^4 dx ,$$

which follows from Hölder's inequality. $X \subset X_3$ by Sobolev's inequality. In order to verify (VI), we use Remark 2. P is locally Lipschitz from X to itself, by Sobolev's inequality. Each statement in (VI) is well-known (see [8] or [7]). (IV) means that

$$\iint \phi^4 dx\, dt + \iint |(m^2-\Delta)^{-1/2}(\partial\phi/\partial t)|^4 dx\, dt < \infty$$

for all free solutions of finite energy, where the integration is over all space-time. Strichartz [16] proved that the first term is finite. If we apply his result to the free solution $\Psi = (m^2-\Delta)^{-1/2}\partial\phi/\partial t$, we get the second term to be finite. Hypothesis (V) means that

$$\int \phi^4 dx + \int |(m^2-\Delta)^{-1/2} \frac{\partial\phi}{\partial t}|^4 dx \le c\, t^{-2}\left\{\int g^{4/3} dx\right\}^3$$

for arbitrary smooth free solutions with $\phi(x,0) = 0$ and $(\partial\phi/\partial t)(x,0) = g(x)$. Just as in the verification of (IV), it suffices to prove the bound on $\int \phi^4 dx$. This bound is the interpolate (halfway) between the energy equality

$$\int \phi^2 dx = \int |(m^2-\Delta)^{-1/2} g|^2 dx$$

and the uniform estimate

$$\sup_x |\phi| \le c\, |t|^{-1} \int (|\nabla g| + |g|) dx .$$

The last estimate follows from the explicit formula for ϕ in terms of g, and an integration by parts. We do not give the details, but refer the reader to the calculation outlined in Appendix B of [6], where a second integration by parts (here unwanted) is carried out.

Theorem 1 may now be applied. The integral and differential equations (in the sense of distributions) are equivalent. Solutions may be continued to arbitrary t . This completes the proof. Note that except for the continuation to times $t > T$, the sign of λ is irrelevant.

General p and n. Replace $\mathbb{R}^3$ by $\mathbb{R}^n$ and the nonlinear term λu^3 by λu^p or $\lambda|u|^{p-1}u$. Its sign does not matter in the application of Theorem 1. Let X be as above, $X_3 = L^{p+1} \oplus W^{-1,p+1}$, $X_1 = \{0\} \oplus L^{1+1/p}$, $d = 2/(p + 1)$. If we assume that $1 < p < 1 + 4/(n - 2)$, then $X \subset X_3$. Strichartz' estimate implies Hypothesis (IV) if in addition $1 + 4/n \leq p$. Hypothesis (V) is equivalent to

$$\int|\phi|^{p+1}dx \leq c\ t^{-2}\{\int|g|^{1+1/p}dx|^p$$

for free solutions ϕ with $\phi(x,0) = 0$, $(\partial\phi/\partial t)(x,0) = g(x)$. We conjecture that this estimate is valid for a broad range of p , in particular for $1 + 4/n \leq p \leq 1 + 4/(n - 1)$.

5. THE NONLINEAR SCHRODINGER EQUATION.

The perturbed and free equations are $i\,u_t - \Delta u + \lambda|u|^{p-1}u = 0$, $i\,v_t - \Delta v = 0$, $x \in \mathbb{R}^n$, $\lambda > 0$. The basic invariants are

$$\int|u|^2dx \quad \text{and} \quad E = \int\{\tfrac{1}{2}|\nabla u|^2 + \frac{\lambda}{p+1}|u|^{p+1}\}dx \quad .$$

We call a solution, for which both invariants are finite, a finite-energy solution.

THEOREM 4. Let $p = 1 + 4/n$. Let u_- be any free solution of finite energy. Then there exists a unique perturbed solution such that

$$u \in C(\mathbb{R}, W^{1,2}(\mathbb{R}^n)) \cap L^{p+1}((-\infty,0)\ ;\ L^{p+1}(\mathbb{R}^n)) \quad ,$$

$$||u(t) - u_-(t)|| \to 0 \quad \text{in the} \quad W^{1,2}(\mathbb{R}^n) \quad \text{norm} \quad ,$$

$$\int|u|^2dx = \int|u_-|^2dx \quad ,$$

$$\int\{|\nabla u|^2 + \frac{2\lambda}{p+1}|u|^{p+1}\}dx = \int|\nabla u_-|^2dx \quad .$$

Proof. Let $X = W^{1,2}$, $X_3 = L^{p+1}$, $X_1 = L^{1+1/p}$. Hypotheses (I), (II), and (III) are easy, as in section 3.

(VI) is proved by Ginibre and Velo [2]; see their Theorem 3.1. We now summarize their proof. The nonlinearity $F(u) = \lambda|u|^{p-1}u$ is approximated (by truncation for instance) by functions $F_\nu(u)$ with the following properties: $\bar{u}F_\nu(u) \geq 0$, $|F_\nu(u)| \leq |F(u)|$, and $F_\nu(u)$ grows slowly enough as $|u| \to \infty$ that the operator $u(\cdot) \to F_\nu(u(\cdot))$ takes X into X . Let P_ν and G_ν be defined naturally in terms of F_ν . By Remark 2, Hypothesis (VI) is valid for each ν . In particular,

$$\frac{1}{2}||u_\nu(t)||_X^2 + G_\nu(u_\nu(t)) = \text{constant}$$

where u_ν is the solution of the F_ν equation with $u_\nu(x,s) = f(x)$ independent of ν . Thus $\{u_\nu\}$ is a bounded sequence in $L^\infty(I,X)$. But $u_\nu \to u$ in L^{p+1} locally in time. So as in section 3 it follows that $u \in L^\infty(I,X)$, u is the weak* limit of $\{u_\nu\}$, and the energy inequality holds. If the roles of s and t in equation (2) are reversed, the energy equality (ii) follows and hence $u \in C(I,X)$.

(IV) follows from the inequality

$$\{\iint|\phi|^{p+1}dx\, dt\}^{1/(p+1)} \leq c\, \{\int|\phi(x,0)|^2dx\}^{1/2} \quad . \tag{11}$$

By dilation (11) can be valid only if $p = 1 + 4/n$. Strichartz [16] proved that it is indeed valid for that critical value of p . To prove (V), we note that $U_o(t)$ is unitary from L^2 onto L^2 and takes L^1 into L^∞ with norm $\leq c\, t^{-n/2}$ for $t > 0$. The latter comes from the explicit formula for $U_o(t)$. By interpolation, $U_o(t)$ takes $L^{1+1/p}$ into L^{p+1} with norm $\leq c\, t^{-d(p)}$ where $d(p) = (np - n)/(2p + 2)$. Only for $p = 1 + 4/n$ is $d = 2/(p + 1)$. This proves (V). Note that the proof for the Klein-Gordon case above is completely analogous to this proof. Solutions may be continued to arbitrary times and so Theorem 4 is proved.

COROLLARY. The wave operator W_- takes the Sobolev space $W^{1,2}$ into itself.

If we assume only that u_- only has L^2 initial data, there are technical complications.

THEOREM 5. Let $p = 1 + 4/n$. Let u_- be any free solution with data in L^2 . Then there exists $T > -\infty$ and a unique perturbed solution u defined for a.e. $t \leq T$ such that

$$u \in L^\infty((-\infty,T), L^2(\mathbb{R}^n)) \cap L^{p+1}((-\infty,T) \times \mathbb{R}^n) ,$$

$$||u_-(t) - u(t)||_{L^2} \to 0 \quad \text{as a.e. } t \to -\infty ,$$

$$||u(t)||_{L^2} = ||u_-(t)||_{L^2} \quad \text{for a.e. } t .$$

Proof. We choose $X = L^2$, $X_3 = L^{p+1}$, $X_1 = L^{1+1/p}$, $G = 0$. We do not have $X \subset X_3$, so we apply Theorem 2. Hypotheses (I)-(V) are verified as before. The almost everywhere version of (VI) is proved just as in Theorem 4.

COROLLARY. W_- takes L^2 into itself.

We conclude this section with some remarks on the critical power $p = 1 + 4/n$. This power makes its appearance basically because of the dilation invariance of the Schrödinger equation. Formally, the Lagrangian is

$$\mathcal{L} = \iint \{-\mathrm{Im}(u_t\bar{u}) + \frac{1}{2}|\nabla u|^2 + \frac{\lambda}{p+1}|u|^{p+1}\}dx\,dt .$$

The quadratic part of $\mathcal{L}$ is invariant under the transformation $u(x,t) \to c^{n/2}u(cx,c^2t)$. The third term is invariant if and only if $p = 1 + 4/n$. By Noether's theorem there must be an associated conservation law. In fact, for any p there is the dilational law

$$\frac{d}{dt}\int \{\frac{1}{4}\mathrm{Im}(r\,u_r\bar{u}) + t(\frac{1}{2}|\nabla u|^2 + \frac{\lambda}{p+1}|u|^{p+1})\}dx + \frac{n}{4}(p-1-4/n)\frac{\lambda}{p+1}\int |u|^{p+1}dx = 0 .$$

Ginibre and Velo [3] discovered the pseudo-conformal law

$$\frac{d}{dt}\int \{\frac{1}{2}|xu - 2it\nabla u|^2 + \frac{4t^2\lambda}{p+1}|u|^{p+1}\}dx + 2nt(p-1-4/n)\frac{\lambda}{p+1}\int |u|^{p+1}dx = 0 .$$

Each of these is a conservation law if and only if $p = 1 + 4/n$. Each of the estimates needed above for free solutions; namely, (11) and

$$\int |\phi|^{p+1} dx \leq c\, t^{-2} \{ \int |\phi(x,0)|^{1+1/p} dx \}^p \quad ;$$

are valid only for this critical value of p, as we can again see by dilation.

The two laws written above have the consequence that if $\lambda < 0$ and $p \geq 1 + 4/n$, solutions with negative energy E blow up in a finite time. Indeed, E is a constant and the two laws above take the form $dE_1/dt \leq 0$ and $dE_2/dt \leq 0$ if the last term in each is dropped. Combining E, E_1 and E_2, we obtain

$$\int r^2 |u|^2 dx \leq 8t^2 E - 16tE_1(0) + 2E_2(0)$$

Thus $\int r^2 |u|^2 dx \to 0$ as $t \to T$, if the solution exists up to time T. Since

$$c = \int |u|^2 dx \leq ||ru||_2 ||u/r||_2 \leq \frac{2}{n-1} ||ru||_2 ||\nabla u||_2$$

we have $||\nabla u||_2 \to \infty$ as $t \to T$. This argument is due to Glassey [4] with modifications of Strauss [14] and Tsutsumi [7]. It proves only non-existence rather than blow-up, unless there is a guarantee that solutions exist up to T. For further discussion of this point, see [1]. If $\lambda > 0$ or if $1 < p < 1 + 4/n$, solutions exist for all time [14].

6. THE WAVE OPERATORS DENSELY DEFINED.

In this section we present an abstraction of part of the work of Ginibre and Velo [3].

THEOREM 6. *Assume* (I), (II) *and* (III). *In* (II) c *is allowed to depend boundedly on* $|f|_3 + |g|_3$. *Omit* (IV) *entirely. Assume* (V) *with* d *allowed to be any constant satisfying* $p^{-1} < d < 1$. *Assume* (VI) *with* $L^{p+1}(I,X_3)$ *replaced by* $C(I,X_3)$. *Let* $f_- \in X \cap X_1$. *Then the conclusion of Theorem 1 is valid if we replace the condition* $u \in L^{p+1}(I,X_3)$ *by* $|u(t)|_3 = O(|t|^{-d})$ *as* $t \to -\infty$.

REMARK. It should be noted that $p^{-1} < 2(p+1)^{-1} < 1$ for $p > 1$. Hence the hypotheses of Theorem 6 are strictly weaker than those of Theorem 1, except that f_- is required to belong to X_1 .

Proof. The proof is similar to that of Theorem 1 and we only indicate the modifications. Let $T < -1$ and $I = (-\infty, T]$. Let

$$||v|| = \sup_I |t|^d |v(t)|_3 \quad , \quad V = \{v \in C(I,X_3) : ||v|| < \infty\} \quad .$$

Throughout the proof, the space $L^{p+1}(I,X_3)$ is replaced by V . Equation (2) clearly makes sense since each term is continuous with values in X_3 . Hence Lemma 1 may be omitted. The analogue of Lemma 2 is valid because

$$|\mathcal{P}u(t)|_3 \le c(||u||)||u||^p \int_{-\infty}^{t} (t-\tau)^{-d} |\tau|^{-dp} d\tau \quad .$$

This integral is $O(|t|^{-d})$ as $t \to -\infty$. Hence $||u|| \le \varepsilon(T)c(||u||)||u||^p$, where $\varepsilon(T) \to 0$ as $T \to -\infty$. As for Lemma 3, we estimate

$$||v_s - u|| < c \sup_{t<T} |t|^d \int_{-\infty}^{s} |t-\tau|^{-d} |\tau|^{-dp} d\tau \quad .$$

This tends to zero as $s \to -\infty$, as we can see by breaking the integral into the part with $|t-\tau| < |t|/2$ and its complement.

Lemma 4 is unchanged except that sequences $\{s_j\}$ are unnecessary. By hypothesis, $|U_o(s)f_-|_3 \to 0$ as $s \to -\infty$. Hence $v_s \to u$ weakly* in $L^\infty(I,X)$ as $s \to -\infty$. Lemmas 5 and 6 are unchanged except that sequences $\{s_j\}$ are unnecessary. As for Lemma 7, we have for $w_s - u$ the same estimate as for $v_s - u$ above. Hence $||w_s - u|| \to 0$ as $s \to -\infty$. The completion of the proof is identical.

In the following application we recover a result of Ginibre and Velo [3].

THEOREM 7. Consider the nonlinear Schrödinger equation, but let

$$\frac{1}{2} + \frac{1}{n} + \left(\frac{1}{4} + \frac{3}{n} + \frac{1}{n^2}\right)^{1/2} = p_o < p < 1 + \frac{4}{n-2} \quad .$$

Then the wave operator W_- takes $W^{1,2} \cap L^{1+1/p}$ into $W^{1,2}$.

Proof. (I), (II), (III) and (VI) are proved just as in Theorem 4, with the same choices of X, X_3 and X_2. From the proof of Theorem 4, we know $U_o(t)$ takes $L^{1+1/p}$ into L^{p+1} with norm $\leq c\, t^{-d(p)}$, where $d(p) = (np-n)/(2p+2)$. It remains to determine the values of p for which $p^{-1} < d(p) < 1$. Thus p_o is the larger root of the polynomial $np^2 - (n+2)p - 2$.

REMARK. A closely related theorem (Theorem 5.3A of [13]) can be used to show that, for the nonlinear Schrödinger equation, W_- is densely defined for all $p > 2 + 2/n$. This overlaps with the condition of Theorem 7. In contrast, a method of Glassey (Theorem 4.3 of [13]) shows that W_- is defined essentially nowhere if $1 < p \leq 1 + 2/n$ (if $n \neq 1$). Naturally, $1 + 2/n < p_o$. Only for p in the interval $(1 + 2/n, p_o]$ is nothing known about the wave operators. This interval is quite small; for $n = 3$, it is $(5/3, 2]$.

REFERENCES

1. J. Ball, Blow-up and nonexistence theorems for nonlinear evolution equations, this volume.

2. J. Ginibre and G. Velo, On a class of nonlinear Schrödinger equations I, to appear.

3. J. Ginibre and G. Velo, On a class of nonlinear Schrödinger equations II, to appear.

4. R. T. Glassey, On the blowing up of solutions to the Cauchy problem for nonlinear Schrödinger equations, J. Math. Phys. 18 (1977), 1794-7.

5. C. S. Morawetz, Time decay for the nonlinear Klein-Gordon equation, Proc. Roy. Soc. A306 (1968), 291-6.

6. C. S. Morawetz and W. A. Strauss, Decay and scattering of solutions of a nonlinear relativistic wave equation, Comm. Pure Appl. Math. 25 (1972), 1-31.

7. M. Reed, Abstract Non-Linear Wave Equations, Lecture Notes in Math. 507, Springer-Verlag, 1976.

8. I. E. Segal, Non-linear semi-groups, Ann. Math. 78 (1963), 339-364.

9. I. E. Segal, Quantization and dispersion for nonlinear relativistic equations, Proc. Conf. Math. Theory Elem. Part., MIT Press, 1966, 79-108.

10. I. E. Segal, Space-time decay for solutions of wave equations, Adv. in Math. 22 (1976), 305-311.

11. E. M. Stein, Singular Integrals and Differentiability Properties of Functions, Princeton U. Press, 1970, page 119.

12. W. A. Strauss, Nonlinear scattering theory, Scattering Theory in Math. Physics, D. Reidel Publ., 1974.

13. W. A. Strauss, Nonlinear invariant wave equations, 1977 Erice lectures, Lecture Notes in Physics, Springer-Verlag.

14. W. A. Strauss, The nonlinear Schrödinger equation, Proc. Conf. Cont. Mech. and PDE, North-Holland, 1978.

15. R. S. Strichartz, A priori estimates for the wave equation and some applications, J. Funct. Anal. 5 (1970), 218-235.

16. R. S. Strichartz, Restrictions of Fourier transforms to quadratic surfaces and decay of solutions of wave equations, Duke Math J. 44 (1977), 705-714.

17. M. Tsutsumi, to appear.

This work was supported in part by NSF Grant MCS75-08827.

Department of Mathematics
Brown University
Providence, Rhode Island 02912

Asymptotic Behavior of Solutions of Evolution Equations

C. M. Dafermos

1. Introduction

We survey here methods of investigation of the asymptotic behavior of solutions of evolution equations, endowed with a dissipative mechanism, based on the study of the structure of the ω-limit set of trajectories of the evolution operator generated by the equation.

The dissipative mechanism usually manifests itself by the presence of a Liapunov functional which is constant on ω-limit sets; the central idea of our approach is to use this information in conjunction with properties of ω-limit sets, such as invariance and minimality. These concepts are only relevant to a special class of evolution operators [14] and, as a result, the applicability of the method is restricted to particular types of evolution equations such as autonomous, periodic, almost periodic, etc. In return one gets results that depend in a delicate fashion on the special evolution character of the equation and cannot be obtained by other techniques. Another advantage of this approach is that it is so simplistic that it only requires quite weak assumptions on the dissipative mechanism. The corresponding drawback is that the deduced information is also weak, never yielding, for example, decay rates of solutions. As a rule, the method is

ISBN 0-12-195250-9

suitable in situations where our information on the dissipative mechanism is somewhat vague or too general. It is noteworthy that many examples of this type arise in continuum physics and control theory.

The present approach originated in the theory of ordinary differential equations. Scattered applications date as far back as the early 1950's (e.g. [10]). It was LaSalle [28,29], however, who saw through the idea and established the method as an "invariance principle" for autonomous and periodic systems of ordinary differential equations. Extensions to other types of nonautonomous differential equations were presented by several authors (e.g. [32,36,1]; for a recent survey and a list of references see [30]). Looking in a different direction, Hale observed that the crucial factor in the application to autonomous systems of ordinary differential equations was the semigroup property of the solution operator. He thus extended [24] the invariance principle to general semigroups on Banach space and gave applications to functional and partial differential equations. In an effort to broaden the class of equations that can be treated by this approach, several authors have contributed extensions of the invariance principle in various directions.

The method will be presented here in an informal fashion, by means of examples, in the hope that this is the best way to reveal the motivation.

In Section 2 we discuss two examples of "wave equations" with weak damping for which the scheme set up by Hale [24] applies. In particular this requires that the Liapunov functional be continuous on phase space. In Sections 3 and 4 we present examples where this condition fails.

The example of Section 3 is a nonlinear wave equation which is visualized as the generator of a dynamical system on the usual "energy space", albeit equipped with the weak topology. This case is treated by an idea of Ball [5] that allows to replace the continuity condition on the "energy" Liapunov functional by a lower semicontinuity condition on the time rate of change of the functional.

Section 4 discusses the case of a hyperbolic conservation law that generates a semigroup on L^1_{loc} space. The lack of continuity of the Liapunov functional is here handled by using the observation [16] that, whenever trajectories are Liapunov stable, a mere lower semicontinuity condition suffices.

The final Section 5 gives a survey of various applications and extensions of these ideas and may serve as a guide to those interested in learning more about the method.

2. Continuous Liapunov Functionals

Consider the problem

$$u_{tt} = \Delta u - a(x)q(u_t) \quad \text{on} \quad [0,\infty) \times \Omega \tag{2.1}$$

$$u = 0 \quad \text{on} \quad [0,\infty) \times \partial\Omega \tag{2.2}$$

$$u = u_0(x) \;,\; u_t = v_0(x) \quad \text{on} \quad \{0\} \times \Omega \tag{2.3}$$

where Ω is an open, bounded, smooth, connected set in R^m; $a(x)$ is smooth and satisfies $a(x) \geq 0$, $x \in \Omega$, $a(x_0) > 0$ for some $x_0 \in \Omega$; $q(v)$ is continuously differentiable, strictly monotone with $q(0) = 0$ and $q'(v)$ bounded on $(-\infty,\infty)$ (this last assumption is only made for the sake of simplicity).

The dissipative mechanism manifests itself by the "energy" integral

$$\frac{d}{dt} E(u,u_t) = -\int_\Omega a(x)q(u_t)u_t dx \leq 0 \tag{2.4}$$

where

$$E(u,v) \overset{\text{def}}{=} \frac{1}{2}\int_\Omega (v^2 + |\nabla u|^2)dx \quad . \tag{2.5}$$

The natural question is whether dissipation drains out the energy and thus drives solutions to zero as $t \to \infty$.

For orientation, consider for a moment the linear case, $q(v) = v$. Decay of $E(u,u_t)$ to zero, as $t \to \infty$, was established in [25] but only after considerable effort. If $a(x) \geq a_0 > 0$, $x \in \Omega$, it is easy to see that the decay is exponential (for an interesting discussion on the rate, see [34].) The work of Rauch and Taylor [35] indicates that we should expect exponential decay even when $a(x)$ vanishes on a subset of Ω, provided that every ray wandering in Ω (reflected at the boundary) crosses recurrently the support of $a(x)$. On the other hand, if there are rays that never intersect the support of $a(x)$, one may construct by the methods of Ralston [33] solutions of the equation that stay for as long as one wishes as close as one pleases to a solution of the undamped wave equation. Thus, in this situation it is impossible to have decay at a uniform rate.

We now return to the general case and show how decay to zero can be established by a very simple argument. We rewrite (2.1) as a first order system

$$\begin{aligned} u_t &= v \\ v_t &= \Delta u - a(x)q(v) \end{aligned} \tag{2.6}$$

which generates a continuous semigroup $T(t)$ on $H_0^1(\Omega) \times L^2(\Omega)$ (as a matter of fact, a contraction semigroup if the space is normed by $E(u,v)^{1/2}$).

At this point we recall some well-known results from elementary topological dynamics.

<u>Proposition 2.1</u>. Let T be a continuous semigroup on a metric space X. If the orbit $\gamma(\chi) \overset{\text{def}}{=} \bigcup_{t\geq 0} T(t)\chi$, through some point $\chi \in X$, is relatively compact in X, then $T(t)\chi \to \omega(\chi)$, $t \to \infty$, where $\omega(\chi)$, the ω-limit set of χ, is a compact, connected subset of X. Furthermore, $\omega(\chi)$ is positive invariant under $T(t)$, i.e., $T(t)\omega(\chi) \subset \omega(\chi)$ for any $t \geq 0$.

<u>Proof</u>. Compactness of $\omega(\chi)$ in X is obvious. Suppose now that A, B are disjoint open sets in X such that $\omega(\chi) \subset A \cup B$, $\omega(\chi) \cap A \neq \phi$, $\omega(\chi) \cap B \neq \phi$. Fix $\varphi \in \omega(\chi) \cap A$, say $\varphi = \lim_{n\to\infty} T(s_n)\chi$, and $\theta \in \omega(\chi) \cap B$, say $\theta = \lim_{n\to\infty} T(\tau_n)\chi$. Using the fact that the trajectory $T(\cdot)\chi$ is continuous on $[0,\infty)$, we determine t_n in the interval defined by s_n and τ_n such that $T(t_n)\chi \in (A \cup B)^c$. Since $\gamma(\chi)$ is relatively compact, we may assume without loss of generality that $T(t_n)\chi \to \psi$. It is clear that $\psi \in \omega(\chi) \cap (A \cup B)^c$ and this contradiction shows that $\omega(\chi)$ is connected.

Finally, to show that $\omega(\chi)$ is positive invariant, note that if $\psi \in \omega(\chi)$, say $\psi = \lim_{n\to\infty} T(t_n)\chi$, and $t \geq 0$, then $T(t)\psi = \lim_{n\to\infty} T(t+t_n)\chi \in \omega(\chi)$. The proof is complete.

Definition 2.1. Let T be a continuous semigroup on X. A Liapunov functional for T is a map $V : X \to R$ such that $V(T(t)\chi) \leq V(\chi)$ for any $\chi \in X$, $t \geq 0$.

Proposition 2.2. Assume that the conditions of Proposition 2.1 hold and that V is a Liapunov functional for T which is continuous on X. Then V is constant on $\omega(\chi)$.

Proof. Let $V_\infty \overset{\text{def}}{=} \lim_{t\to\infty} V(T(t)\chi)$. Pick any $\psi \in \omega(\chi)$, say $\psi = \lim_{n\to\infty} T(t_n)\chi$. Since V is continuous on X, $V(\psi) = \lim_{n\to\infty} V(T(t_n)\chi) = V_\infty$. The proof is complete.

We now apply the above results to the semigroup $T(t)$ generated by (2.6). We assume that the initial data lie in $[H^2(\Omega) \cap H_0^1(\Omega)] \times H_0^1(\Omega)$ (the domain of the generator of $T(t)$) and prove that the orbit $\gamma((u_0,v_0))$ is relatively compact in $H_0^1(\Omega) \times L^2(\Omega)$. Differentiating (2.1) with respect to t,

$$v_{tt} = \Delta v - a(x)q'(v)v_t \quad . \tag{2.7}$$

The "energy" integral

$$\frac{d}{dt} E(v,v_t) = -\int_\Omega a(x)q'(v)v_t^2 dx \leq 0 \tag{2.8}$$

of (2.7) shows that $\{v(t,\cdot)\,|\,t \geq 0\}$ is bounded in $H_0^1(\Omega)$, and is therefore relatively compact in $L^2(\Omega)$, and that $\{v_t(t,\cdot)\,|\,t \geq 0\}$ is bounded in $L^2(\Omega)$. It follows from (2.1) that $\{\Delta u(t,\cdot)\,|\,t \geq 0\}$ is bounded in $L^2(\Omega)$ and, consequently, $\{u(t,\cdot)\,|\,t \geq 0\}$ is relatively compact in $H_0^1(\Omega)$.

Thus

$$T(t)(u_o,v_o) \xrightarrow{H_0^1 \times L^2} \omega((u_0,v_o)),\ t \to \infty \quad . \tag{2.9}$$

By (2.4) $E(u,v)$ is a Liapunov functional for $T(t)$ which is continuous on $H_0^1(\Omega) \times L^2(\Omega)$ so that (Proposition 2.2)

$E(u,v)$ is constant on $\omega((u_0,v_0))$. Using the positive invariance property of $\omega((u_0,v_0))$ (Proposition 2.1) in conjunction with (2.4) and the properties of $a(x)$ and $q(v)$, we arrive at the following conclusion: If $(\hat{u}(t,\cdot), \hat{v}(t,\cdot)) = T(t)(\hat{u}_0,\hat{v}_0)$ with $(\hat{u}_0,\hat{v}_0) \in \omega((u_0,v_0))$, then

$$\hat{v}(t,x) = 0, \quad \text{for all} \quad t \geq 0, \ x \in \operatorname{spt} a(\cdot) \quad . \tag{2.10}$$

In particular, $\hat{u}(t,\cdot)$ becomes a solution of the undamped wave equation so that we have the representation

$$(\hat{u}(t,x), \hat{v}(t,x)) = \operatorname{Re} \sum_n \exp(i\lambda_n t) w_n(x)(1,i\lambda_n) \tag{2.11}$$

where

$$\Delta w_n + \lambda_n^2 w_n = 0 \qquad \text{on} \quad \Omega \tag{2.12}$$

$$w_n = 0 \qquad \text{on} \quad \partial\Omega \quad . \tag{2.13}$$

Substituting $\hat{v}(t,x)$ from (2.11) into (2.10) and using the elementary properties of almost periodic functions we deduce that $w_n(x) = 0$, $x \in \operatorname{spt} a(\cdot)$. Since solutions of (2.12) are analytic in Ω and Ω is connected, it follows that $w_n(x) = 0$, $x \in \Omega$, so that $(\hat{u}(t,x), \hat{v}(t,x)) = 0$, $t \geq 0, x \in \Omega$. Thus solutions of (2.6) through $(u_0,v_0) \in [H^2(\Omega) \cap H_0^1(\Omega)] \times H_0^1(\Omega)$ tend to zero as $t \to \infty$. Using a simple completion argument we extend the result to the case of initial data in $H_0^1(\Omega) \times L^2(\Omega)$, that is,

Proposition 2.3. Let $u(t,x)$ be the solution of (2.1)-(2.3) with $(u_0,v_0) \in H_0^1(\Omega) \times L^2(\Omega)$. Then

$$(u(t,\cdot), u_t(t,\cdot)) \xrightarrow{H_0^1 \times L^2} 0, \ t \to \infty \quad . \tag{2.14}$$

We should emphasize here that although the procedure followed in the proof of Proposition 2.3 appears quite simplistic it is not trivial since it depends in a delicate fashion on the autonomous character of system (2.6). For example one can exhibit functions $a(t,x)$ with $a(t,x) \geq a_0 > 0$ for $t \geq 0$, $x \in \Omega$, such that solutions of

(2.15) $$u_{tt} = \Delta u - a(t,x)u_t$$

with boundary conditions (2.2) do not decay to zero as $t \to \infty$. We will return to this point in Section 5.

We close this section with a brief discussion of another example where the dissipative mechanism is even weaker and, as a result, not always as effective. We consider the problem

(2.16) $$\vec{u}_{tt} = \Delta\vec{u} + \nabla[a(x)q(\nabla\cdot\vec{u}_t)] \quad \text{on} \quad [0,\infty) \times \Omega$$

(2.17) $$\vec{u} = \vec{0} \quad \text{on} \quad [0,\infty) \times \partial\Omega$$

(2.18) $$\vec{u} = \vec{u}_0(x),\ \vec{u}_t = \vec{v}_0(x) \quad \text{on} \quad \{0\} \times \Omega$$

where $\vec{u}$ is an m-vector field while Ω, $a(x)$ and $q(v)$ are as in the previous example.

In the place of (2.4) we now have the "energy" integral

(2.19) $$\frac{d}{dt} E(\vec{u},\vec{u}_t) = - \int_\Omega a(x)q(\nabla\cdot\vec{u}_t)(\nabla\cdot u_t)dx \leq 0$$

where

(2.20) $$E(\vec{u},\vec{v}) \overset{\text{def}}{=} \frac{1}{2}\int_\Omega (|\vec{v}|^2 + |\nabla\vec{u}|^2)dx \quad .$$

Following the steps in our discussion for the previous example we arrive here at the following

<u>Proposition 2.4</u>. If $\vec{u}(t,x)$ is the solution of (2.16)-(2.18), a necessary and sufficient condition that

(2.21) $$(\vec{u}(t,\cdot),\ \vec{u}_t(t,\cdot)) \xrightarrow{[H_0^1]^m \times [L^2]^m} 0,\ t \to \infty \quad ,$$

for every $(\vec{u}_0,\vec{v}_0) \in [H_0^1(\Omega)]^m \times [L^2(\Omega)]^m$, is that zero is the only solution of the overdetermined system

(2.22) $$\Delta\vec{w} + \lambda^2\vec{w} = \vec{0} \quad \text{on} \quad \Omega$$

(2.23) $$\nabla\cdot\vec{w} = 0 \quad \text{on} \quad \Omega$$

(2.24) $$\vec{w} = \vec{0} \quad \text{on} \quad \partial\Omega \quad .$$

System (2.22) - (2.24) also arises in connection to linear isotropic thermoelasticity theory and is studied in [19]. It is demonstrated there that (2.22) - (2.24) admits nontrivial solutions in certain domains Ω (e.g. the

2-circle). On the other hand it is also shown that zero is the only solution of (2.22) - (2.24) whenever all eigenvalues of the eigenvalue problem (2.12) - (2.13) are simple. As shown by D. Henry (unpublished) this last condition holds generically within the class of smooth Ω. We may thus conclude that solutions of (2.16) - (2.18) decay to zero, as $t \to \infty$, generically with respect to Ω.

3. Use of the Weak Topology

We exhibit here an example for which the approach of Section 2 fails so that a modified procedure has to be developed. We consider the problem

$$u_{tt} = \Delta u + u^3 - a(x)u_t \quad \text{on} \quad [0,\infty) \times \Omega \tag{3.1}$$

$$u = 0 \quad \text{on} \quad [0,\infty) \times \partial\Omega \tag{3.2}$$

$$u = u_0(x),\ u_t = v_0(x) \quad \text{on} \quad \{0\} \times \Omega \tag{3.3}$$

where Ω is an open, bounded, smooth subset of R^3; $a(x)$ is smooth and $a(x) > 0$, $x \in \Omega$. Setting

$$E(u,v) \overset{\text{def}}{=} \frac{1}{2} \int_\Omega (v^2 + |\nabla u|^2 - \frac{1}{2} u^4)dx \quad , \tag{3.4}$$

we have the "energy" integral

$$\frac{d}{dt} E(u,u_t) = - \int_\Omega a(x)u_t^2 dx \leq 0 \tag{3.5}$$

which indicates the presence of a dissipative mechanism. The fact that $E(u,v)$ is not positive definite induces the possibility of equilibrium states other than the trivial one $(u = 0,\ u_t = 0)$ and this makes the problem of asymptotic behavior particularly interesting.

We rewrite (3.1) as the first order system

$$\begin{aligned} u_t &= v \\ v_t &= \Delta u + u^3 - a(x)v \quad . \end{aligned} \tag{3.6}$$

For any assigned initial data $(u_0,v_0) \in H_0^1(\Omega) \times L^2(\Omega)$ there is a unique local solution of (3.6) with values in $H_0^1(\Omega) \times L^2(\Omega)$. However, due to the indefiniteness of the "energy" functional (3.4), some of these solutions escape in finite time (see [6]). Nevertheless, there are solutions (in particular those with smooth, small initial data) that are defined for all $t \geq 0$ and are bounded in $H_0^1(\Omega) \times L^2(\Omega)$. It is the asymptotic behavior of these solutions that we are planning to determine here.

In attempting to employ in the present case the procedure of Section 2, one is faced with the obstacle of showing that the range of solutions of (3.6) is relatively compact in $H_0^1(\Omega) \times L^2(\Omega)$. The method used in Section 2 does not apply here (if one differentiates (3.1) with respect to t, the "nondissipative" term $3u^2u_t$ emerges). One way out is to employ, instead, the weak topology of $H_0^1(\Omega) \times L^2(\Omega)$ in which case boundedness suffices for relative compactness. We thus proceed along the following lines: Starting with some fixed solution of (3.6) which is bounded in $H_0^1(\Omega) \times L^2(\Omega)$, we denote by X the weak closure of its range. Thus X, equipped with the weak topology of $H_0^1(\Omega) \times L^2(\Omega)$, is a compact metric space.

It can be shown [5] that solutions of (3.6) depend continuously on initial data, relative to the weak topology of $H_0^1(\Omega) \times L^2(\Omega)$, and this implies that (3.6) generates a continuous semigroup $T(t)$ on X. In particular, Proposition 2.1 applies.

The next step is to employ the Liapunov functional $E(u,v)$ defined by (3.4). Here one has to pay the price of having adopted the weak topology of $H_0^1(\Omega) \times L^2(\Omega)$ because $E(u,v)$ is not continuous relative to this topology and, as a result, Proposition 2.2 does not apply. In order to overcome this difficulty, Ball [5] establishes the following

Proposition 3.1. Assume that the conditions of Proposition 2.1 hold and that V is a Liapunov functional for T, bounded on compact subsets of X, and such that, for any fixed $t \geq 0$, the map $\psi \mapsto V(\psi) - V(T(t)\psi)$ is lower semicontinuous on X. Then, if $\psi \in \omega(\chi)$,

$$V(T(t)\psi) = V(\psi), \quad \text{for all } t \geq 0 \quad . \tag{3.7}$$

Proof. Let $\psi \in \omega(\chi)$, say $\psi = \lim_{n\to\infty} T(t_n)\chi$. Then, for $t \geq 0$, $0 \leq V(\psi) - V(T(t)\psi) \leq \liminf_{n\to\infty}[V(T(t_n)\chi) - V(T(t+t_n)\chi)] = 0$. The proof is complete.

Returning to our example, we obtain from (3.5)

$$E(u(0,\cdot),\, v(0,\cdot)) - E(u(t,\cdot),\, v(t,\cdot)) = \int_0^t \int_\Omega a(x) v^2(\tau,x)\,dx\,d\tau \tag{3.8}$$

which shows that the Liapunov functional $E(u,v)$ satisfies the assumptions of Proposition 3.1. Applying this proposition in conjunction with Proposition 2.1 and recalling that $a(x) > 0$, $x \in \Omega$, we arrive at the following

Proposition 3.2. Let $u(t,x)$ be a solution of (3.1) - (3.3) such that $(u(t,\cdot),\, u_t(t,\cdot))$ is bounded in $H_0^1(\Omega) \times L^2(\Omega)$. Then any sequence $\{t_n\}$ in $[0,\infty)$, with $t_n \to \infty$, contains a subsequence, denoted again by $\{t_n\}$, such that

$$(u(t_n,\cdot),\, u_t(t_n,\cdot)) \xrightarrow[\text{weakly}]{H_0^1 \times L^2} (\hat{u}(\cdot),0) \tag{3.9}$$

where $\hat{u}(x)$ is some solution of the boundary value problem

$$\Delta\hat{u} + \hat{u}^3 = 0 \qquad \text{on } \Omega \tag{3.10}$$

$$\hat{u} = 0 \qquad \text{on } \partial\Omega\ . \tag{3.11}$$

The above information is weak on several counts. In the first place, the weak convergence in (3.9) does not reveal whether the entire energy loss is due to the dissipative mechanism. Furthermore, since (3.10) - (3.11) may have many solutions, the ω-limit set of a single orbit may contain an infinite number of states corresponding to different "energy levels" of the Lagrangian

$$W(\hat{u}) \overset{\text{def}}{=} \frac{1}{2}\int_\Omega \left(|\nabla\hat{u}|^2 - \frac{1}{2}\hat{u}^4\right)dx = E(\hat{u},0) \tag{3.12}$$

(note that, in contrast to Proposition 2.2, Proposition 3.1 only asserts that in the ω-limit set the Liapunov functional is constant just along trajectories.)

One can draw stronger conclusions when the damping is strong, in the sense

$$a(x) \geq a_0 > 0 \quad , \quad x \in \Omega\ . \tag{3.13}$$

In this case one may attempt to show that orbits of $T(t)$ are actually relatively compact with respect to the strong topology of $H_0^1(\Omega) \times L^2(\Omega)$ by applying the method of Webb [46]: In (3.6) one visualizes the term u^3 as a perturbation of a linear system which generates, on account of (3.13), an exponentially asymptotically stable semigroup $S(t)$. One then tries to deduce relative compactness of the ordit from the variation of parameters formula

$$T(t)(u_0,v_0) = S(t)(u_0,v_0) + \int_0^t S(t-\tau)(0,u^3)d\tau \quad . \tag{3.14}$$

As noted by J. Ball, the above approach would work if we had in (3.6) $|u|^{\gamma-1}u$, $1 \leq \gamma < 3$, in the place of u^3, but fails in the present case (the reason being that the injection of $H_0^1(\Omega)$ in $L^{2\gamma}(\Omega)$ is compact for $1 \leq \gamma < 3$ but not for $\gamma = 3$).

We now give a proof of strong convergence by an argument in the spirit of [5]. A simple calculation establishes the following identity for solutions of (3.6):

$$\begin{aligned} &\frac{d}{dt}\int_\Omega uv dx - 2\int_\Omega v^2 dx = \\ &- 2E(u,v) + \frac{1}{2}\int_\Omega u^4 dx - \int_\Omega a(x)uv dx \quad . \end{aligned} \tag{3.15}$$

On account of (3.5) and (3.13), $v(t,x) \in L^2([0,\infty) \times \Omega)$ so that the integral of the left-hand side of (3.15) over any t-interval is uniformly bounded. On the other hand, by (3.9) and the embedding theorem, the three terms on the right-hand side of (3.15) tend, as $t \to \infty$, to $-2E_\infty$, $\frac{1}{2}\int_\Omega \hat{u}^4 dx$ and 0, respectively. It follows that

$$E_\infty = \frac{1}{4}\int_\Omega \hat{u}^4 dx \quad . \tag{3.16}$$

By virtue of (3.10), (3.11) and (3.12), (3.16) yields $E_\infty = W(\hat{u}) = E(\hat{u},0)$. In particular,

$$\|T(t_n)(u_0,v_0)\|_{H_0^1 \times L^2} \to \|(\hat{u},0)\|_{H_0^1 \times L^2} \tag{3.17}$$

so that the convergence in (3.9) becomes strong. It is still possible that $\omega((u_0,v_0))$ contain many states; however, they all have to be at the same energy level. The connectedness of $\omega((u_0,v_0))$ (Proposition 2.1) narrows down the possibilities even further. We thus have

<u>Proposition 3.3.</u> Under the assumption (3.13), let $u(t,x)$ be a solution of (3.1) - (3.3) such that $(u(t,\cdot), u_t(t,\cdot))$ is bounded in $H_0^1(\Omega) \times L^2(\Omega)$. Suppose, further, that for each $W_0 \in (-\infty,\infty)$ there is an at most finite number of solutions $\hat{u}(x)$ of (3.10) - (3.11) with $W(\hat{u}) = W_0$. Then

$$(3.18) \qquad (u(t,\cdot), u_t(t,\cdot)) \xrightarrow{H_0^1 \times L^2} (\hat{u}(\cdot),0), \ t \to \infty \quad ,$$

where $\hat{u}(x)$ is a solution of (3.10) - (3.11).

4. <u>Lower Semicontinuous Liapunov Functionals</u>

We now discuss another example for which the method of Section 2 does not apply and we thus have to use an alternative argument.

We consider the conservation law

$$(4.1) \qquad u_t + f(u)_x = 0 \quad , \quad t \geq 0, -\infty < x < \infty \quad .$$

We are interested in solutions of the initial value problem for (4.1) in the class of bounded measurable functions. We recall that, in order to secure uniqueness of solutions within this class, one has to adopt certain solution admissibility criteria (for three different approaches see [27, 26, 12]). The fundamental estimates are as follows: If $u(t,x)$, $v(t,x)$ are two admissible solutions of (4.1), then, for any $-\infty < a < b < \infty$,

$$(4.2) \qquad \int_a^b |u(t,x) - v(t,x)|dx \leq \int_{a-Mt}^{b+Mt} |u(0,x) - v(0,x)|\, dx$$

where $M = \max\{\|f'(u)\|_{L^\infty}, \|f'(v)\|_{L^\infty}\}$. Furthermore, if $u(0,x) \leq v(0,x)$, $x \in (-\infty,\infty)$, then

$$(4.3) \qquad u(t,x) \leq v(t,x), \quad \text{for all} \quad t \geq 0, -\infty < x < \infty \quad .$$

In particular,

$$\text{(4.4)} \quad \begin{aligned} &\operatorname{ess}_x\sup u(t,x) \le \operatorname{ess}_x\sup u(0,x) \quad , \quad \text{for all} \quad t \ge 0 \\ &\operatorname{ess}_x\inf u(t,x) \ge \operatorname{ess}_x\inf u(0,x) \quad , \quad \text{for all} \quad t \ge 0 \ . \end{aligned}$$

When $f(u)$ is linear, (4.4) hold as equalities so that solutions do not decay. However, if $f(u)$ is nonlinear, the possibility of wave interactions induces a dissipative mechanism. The dissipation is strongest when (4.1) is "genuinely nonlinear", i.e., $f''(u) \neq 0$. In this case one establishes decay of solutions with uniform rates [31]. Decay with uniform rates can also be established [23, 21] even when $f''(u)$ is allowed to vanish at isolated points albeit in a controlled fashion. One observes that the rates of decay get "lower" as $f(u)$ gets "flatter" at points of inflexion. Here we study the asymptotic behavior of solutions under the sole assumption that the set of zeros of $f''(u)$ has no accumulation point on the real line. We do not impose any restrictions on the "flatness" of $f(u)$ at its points of inflexion and, of course, we do not expect decay at a uniform rate.

For definiteness, we will consider initial data in the set X of functions that are essentially bounded and L^1-almost periodic (for the periodic case see [17]). Functions $v(\cdot) \in X$ will be equipped with the norm

$$\text{(4.5)} \qquad \|v\| \overset{\text{def}}{=} \lim_{L\to\infty} \frac{1}{2L} \int_{-L}^{L} |v(x)|dx \ .$$

Estimates (4.2), (4.3) indicate that (4.1) generates a contraction semigroup $T(t)$ on X. The same estimates also yield that every orbit of $T(t)$ consists of uniformly L^1-almost periodic functions and is, therefore, relatively compact in X. Proposition 2.1 thus applies.

It is easy to construct Liapunov functionals that are continuous on orbits of $T(t)$. For example, if $\eta(u)$ is any convex function (entropy) it can be shown that

$$\text{(4.6)} \qquad \lim_{L\to\infty} \frac{1}{2L} \int_{-L}^{L} \eta(v(x))dx$$

is such a functional. However, the idea of applying Proposition 2.2 to one of these functionals does not lead to any useful conclusions because it is very difficult to characterize positive invariant sets on which (4.6) is constant.

On account of (4.4),

$$V(v) \overset{\text{def}}{=} \operatorname{ess}_x \sup v(x) \tag{4.7}$$

is a Liapunov functional for $T(t)$. In view of the ordering principle (4.3), functional (4.7) can be very effective. The difficulty is that V is not continuous but merely lower semicontinuous on X. We now proceed to show that in the present case semicontinuity of V suffices because $T(t)$ is a contraction semigroup on X.

Definition 4.1. Let T be a semigroup on a metric space X (metric ρ). The trajectory $T(\cdot)\chi$ through a point $\chi \in X$ is called Liapunov stable if for any $\varepsilon > 0$ there is $\delta(\varepsilon)$ such that $\rho(\psi,\chi) < \delta(\varepsilon)$ implies $\rho(T(t)\psi, T(t)\chi) < \varepsilon$ for all $t \geq 0$.

Proposition 4.1. Let T be a continuous semigroup on a metric space X (metric ρ). Assume that the orbit $\gamma(\chi)$ through some point $\chi \in X$ is relatively compact in X and that the trajectory through any point of $\overline{\gamma(\chi)}$ is Liapunov stable. If V is a Liapunov functional for T which is lower semicontinuous on X, then V is constant on $\omega(\chi)$.

Proof. Let $\varphi, \psi \in \omega(\chi)$, say $\varphi = \lim_{m\to\infty} T(s_m)\chi$, $\psi = \lim_{n\to\infty} T(t_n)\chi$, and suppose $V(\varphi) \leq V(\psi)$. We construct a subsequence $\{t_m\}$ of $\{t_n\}$ such that $\tau_m \overset{\text{def}}{=} t_m - s_m > 0$. Then

$$\rho(T(\tau_m)\varphi, \psi) \leq \rho(T(\tau_m)\varphi,\ T(\tau_m)T(s_m)\chi) + \rho(T(t_m)\chi, \psi)$$

which shows that $T(\tau_m)\varphi \to \psi$, $m \to \infty$. It follows that $V(\psi) \leq \liminf_{m\to\infty} V(T(\tau_m)\varphi) \leq V(\varphi)$. Thus $V(\varphi) = V(\psi)$. The proof is complete.

Returning to the semigroup $T(t)$ on X generated by (4.1) we have, on account of Propositions 2.1 and 4.1, that if $u_0(x)$ lies in the ω-limit set of an orbit and if $u(t,x)$ denotes the admissible solution of (4.1) with initial condition $u(0,x) = u_0(x)$, then

(4.8) $$\text{ess}_x\sup u(t,x) = \text{ess}_x\sup u_0(x) \overset{\text{def}}{=} W, \quad \text{for all} \quad t \geq 0 .$$

We claim that $u_0(x) = W$ almost everywhere on $(-\infty,\infty)$. Indeed, if this is not true, we can find $\varepsilon > 0$ such that $f''(u)$ does not vanish in the interval $(W-\varepsilon, W)$ and, at the same time, the set $\{x \in (-\infty,\infty) | u_0(x) \leq W-\varepsilon\}$ has positive measure. Consider the function

(4.9) $$v_0(x) \overset{\text{def}}{=} \begin{cases} u_0(x) & \text{if} \quad u_0(x) > W-\varepsilon \\ W-\varepsilon & \text{if} \quad u_0(x) \leq W-\varepsilon \end{cases} .$$

Note that $v_0(x) \in X$ and that

(4.10) $$\overline{v}_0 \overset{\text{def}}{=} \lim_{L\to\infty} \frac{1}{2L} \int_{-L}^{L} v_0(x)\,dx \in (W-\varepsilon, W).$$

Let $v(t,x)$ be the admissible solution of (4.1) with initial condition $v(0,x) = v_0(x)$. By (4.3), $u(t,x) \leq v(t,x)$ for all $t \geq 0$, $x \in (-\infty,\infty)$. The range of $v(t,x)$ is contained in the interval $(W-\varepsilon, W)$ on which $f''(u) \neq 0$. Then it is known [31,21] that $v(t,x) \to \overline{v}_0$, $t \to \infty$, uniformly in $x \in (-\infty,\infty)$. Therefore, for t sufficiently large, $\text{ess}_x\sup u(t,x) < W$ which is a contradiction to (4.8). We have thus shown that $u_0(x)$ is constant on $(-\infty,\infty)$. The value of the constant is determined easily by the observation that the x-mean of solutions of (4.1) is conserved in time. We summarize the result in the following

Proposition 4.2. Let $u(t,x)$ be the admissible solution of (4.1) with initial datum $u(0,x)$ which is bounded, L^1-almost periodic and has mean zero, i.e.,

(4.11) $$\lim_{L\to\infty} \frac{1}{2L} \int_{-L}^{L} u(0,x)\,dx = 0 .$$

Then

(4.12) $$\lim_{L\to\infty} \frac{1}{2L} \int_{-L}^{L} |u(t,x)|\,dx \to 0, \quad t \to \infty .$$

5. Applications and Extensions

A natural question is whether the ideas described in the previous sections apply to nonautonomous evolution equations as well. The issue rests on whether Proposition 2.1 can be extended from semigroups to evolution operators, that is families of continuous maps $U(t,\tau):X \to X$, $t \in (-\infty,\infty)$, $\tau \in [0,\infty)$, X a metric space, satisfying

$$(5.1) \qquad U(t,0) = \text{identity}, \quad t \in (-\infty,\infty) \quad ,$$

$$(5.2) \quad U(t,\sigma+\tau) = U(t+\sigma,\tau)U(t,\sigma), \quad t \in (-\infty,\infty), \ \sigma,\tau \in [0,\infty) \quad .$$

If U is an evolution operator and $s \in (-\infty,\infty)$, the s-translate of U is the evolution operator U_s defined by

$$(5.3) \qquad U_s(t,\tau) = U(t+s,\tau), \ t \in (-\infty,\infty), \ \tau \in [0,\infty) \quad .$$

Guided by the experience accumulated in the theory of ordinary differential equations [28,29,32,36,30,1], it is possible to identify the class of evolution operators U for which analogs of Proposition 2.1 hold. This is the class of <u>compact processes</u> [14] characterized by the property that the family $\{U_s | s \in [0,\infty)\}$ of right translates is equicontinuous, for fixed $t \in (-\infty,\infty)$, $\tau \in [0,\infty)$, and sequentially relatively compact in the pointwise topology. Examples of compact processes are semigroups (for which $U_s = U$ for every $s \in (-\infty,\infty)$), periodic processes (in which $U_L = U$ for some period $L > 0$), almost periodic processes, asymptotically almost periodic processes, etc. (see [37,14,18]). The theory of compact processes yields, for example, that the assertion of Proposition 2.3 holds also for solutions of the nonautonomous equation

$$(5.4) \qquad u_{tt} = \Delta u - \beta(t)a(x)q(u_t) + f(t,x)$$

where $a(x) \geq 0$, $a(x_0) > 0$, $q(0) = 0$, $0 < \mu \leq q'(v) \leq M$, $\beta(t) > 0$ is C^1 almost periodic and $f(t,\cdot) \in L^1([0,\infty);L^2(\Omega))$.

In the same set-up one can study the problem of existence and stability of almost periodic solutions to almost periodic evolution equations [20]. For example, it can be shown that (5.4) - (2.2), with $a(x) \geq a_0 > 0$ for all $x \in \Omega$

and $f(t,\cdot)$ almost periodic, possesses a unique almost periodic solution which attracts asymptotically every other solution.

It is also possible to identify the class of evolution operators, called uniform processes [16,18], for which an analog of Proposition 4.1 holds and, therefore, lower semicontinuity of Liapunov functionals suffices. Other applications of lower semicontinuous Liapunov functionals are indicated in [44].

The need to handle specific examples arising in the applications has also led to extensions in different directions. For example, if the initial value problem for the evolution equation may have more than one solution, the semigroup (or evolution operator) is not well-defined. Ball [5] proposes a set-up of the problem that overcomes this obstacle and gives references to related work by other authors. Another observation is that the defining property of Liapunov functionals, to be decreasing along trajectories (Definition 2.1), is only used, in the proof of Proposition 2.2, to infer that $V(T(t)\chi)$ tends to a limit, as $t \to \infty$, and, in the proof of Proposition 3.1, to deduce that $V(T(t_n)\chi)-V(T(t+t_n)\chi)$ tends to zero, as $t_n \to \infty$. It is clear, however, that such conclusions may be drawn under weaker assumptions on V. This remark leads to useful extensions of the concept of a Liapunov functional [5]. Actually [5] contains a wealth of ideas and should be studied by those interested in the methods discussed here. For other relevant theoretical developments see [38,22,45,43,3,4].

We close this section with a list of references pertaining to applications. Weak dissipative mechanisms arise quite often in continuum physics, in the form of viscous, diffusive or thermal damping. Asymptotic behavior results by the present method have been obtained for linear thermoelasticity [42,19], viscoelasticity of the Boltzmann type [13,19], beam theory [2], linearized theory of simple fluids [41] and the theory of mixtures [19]. A number of natural applications of the method in control theory have also been presented [39,40, 9]. For other applications see [24,15,11,7,8].

REFERENCES

[1] Artstein, Z., The limiting equations of nonautonomous ordinary differential equations. J. Diff. Eqs. 25 (1977), 184-202.

[2] Ball, J. M., Stability analysis for an extensible beam. J. Diff. Eqs. 14 (1973), 399-418.

[3] Ball, J. M., Measurability and continuity conditions for nonlinear evolutionary processes. Proc. Amer. Math. Soc. 55 (1976), 353-358.

[4] Ball, J. M., Measurability and continuity conditions for evolutionary processes. "Dynamical Systems", Vol. II, L. Cesari, J. K. Hale & J. P. LaSalle eds., Academic Press, New York 1976.

[5] Ball, J. M., On the asymptotic behaviour of generalised processes, with applications to nonlinear evolution equations. J. Diff. Eqs. (to appear).

[6] Ball, J. M., Finite time blow-up in nonlinear problems. This Volume.

[7] Ball, J. M. & Peletier, L. A., Stabilization of concentration profiles in catalyst particles. J. Diff. Eqs. 20 (1976), 356-368.

[8] Ball, J. M. & Peletier, L. A., Global attraction for the one-dimensional heat equation with nonlinear time-dependent boundary conditions. Arch. Rat. Mech. Anal. 65 (1977), 193-201.

[9] Ball, J. M. & Slemrod, M., Feedback stabilization of distributed semilinear control systems. (Preprint).

[10] Barbashin, E. A. & Krasovskii, N. N., Stability of motion in the large. Dokl. Akad. Nauk SSSR 86 (1952), 453-456.

[11] Chafee, N. & Infante, E. F., A bifurcation problem for a nonlinear partial differential equation of parabolic type. Applicable Anal. 4 (1974), 17-37.

[12] Crandall, M. G., The semigroup approach to first order quasilinear equations in several space variables. Israel J. Math. 12 (1972), 108-132.

[13] Dafermos, C. M., Asymptotic stability in viscoelasticity. Arch. Rat. Mech. Anal. 37 (1970), 297-308.

[14] Dafermos, C. M., An invariance principle for compact processes. J. Diff. Eqs. 9 (1971), 239-252.

[15] Dafermos, C. M., Applications of the invariance principle for compact processes. J. Diff. Eqs. 9 (1971), 291-299.

[16] Dafermos, C. M., Uniform processes and semicontinuous Liapunov functionals. J. Diff. Eqs. 11 (1972), 401-415.

[17] Dafermos, C. M., Asymptotic behavior of solutions of a hyperbolic conservation law. J. Diff. Eqs. 11 (1972), 416-424.

[18] Dafermos, C. M., Semiflows generated by compact and uniform processes. Math. Syst. Theory 8 (1974), 142-149.

[19] Dafermos, C. M., Contraction semigroups and trend to equilibrium in continuum mechanics. Springer Lecture Notes in Math. No. 503 (1976), 295-306.

[20] Dafermos, C. M., Almost periodic processes and almost periodic solutions of evolution equations. "Dynamical Systems", A. Bednarek & L. Cesari eds., pp. 43-57, Academic Press, New York 1977.

[21] Dafermos, C. M., Characteristics in hyperbolic conservation laws. A study of the structure and the asymptotic behavior of solutions. "Nonlinear Analysis and Mechanics", Heriot-Watt Symposium, Vol. I, Pitman, London 1977.

[22] Dafermos, C. M. & Slemrod, M., Asymptotic behavior of nonlinear contraction semigroups. J. Funct. Anal. 13 (1973), 97-106.

[23] Greenberg, J. M. & Tong, D. M., Decay of periodic solutions of $\partial u/\partial t+\partial f(u)/\partial x = 0$. J. Math. Anal. Appl. 43 (1973), 56-71.

[24] Hale, J. K., Dynamical systems and stability. J. Math. Anal. Appl. 26 (1969), 39-59.

[25] Iwasaki, N., Local decay of solutions for symmetric hyperbolic systems with dissipative and coercive boundary conditions in exterior domains. Publ. Res. Inst. Math. Sci., Kyoto Univ., 5 (1969), 193-218.

[26] Keyfitz, B., Solutions with shocks. An example of an L^1-contraction semigroup. Comm. Pure Appl. Math. 24 (1971), 125-132.

[27] Kruzkov, S. N., First order quasilinear equations in several independent variables. Mat. Sbornik (N.S.) 81 (123) (1970), 228-255. English translation: Math USSR-Sbornik 10 (1970), 217-243.

[28] LaSalle, J. P., The extent of asymptotic stability. Proc. Nat. Acad. Sci. USA, 46 (1960), 363-365.

[29] LaSalle, J. P., Asymptotic stability criteria. Proc. Symp. Appl. Math. 13 (1962), 299-307. Amer. Math. Soc., Providence.

[30] LaSalle, J. P., (with an appendix by Z. Artstein), The stability of dynamical systems. SIAM regional conference series in Applied Math. No. 25 (1976).

[31] Lax, P. D., Hyperbolic systems of conservation laws II. Comm. Pure Appl. Math. 10 (1957), 537-566.

[32] Miller, R. K., Asymptotic behavior of solutions of non-linear differential equations. Trans. Amer. Math. Soc. 115 (1965), 400-416.

[33] Ralston, J. V., Solutions of the wave equation with localized energy. Comm. Pure Appl. Math. 22 (1969), 807-323.

[34] Rauch, J., Qualitative behavior of dissipative wave equations on bounded domains. Arch. Rat. Mech. Anal. 62 (1976), 77-85.

[35] Rauch, J. & Taylor, M., Decay of solutions to non dissipative hyperbolic systems on compact manifolds. Comm. Pure Appl. Math. 28 (1975), 501-523.

[36] Sell, G. R., Nonautonomous differential equations and topological dynamics, I, II. Trans. Amer. Math. Soc. 127 (1967), 241-283.

[37] Slemrod, M., Asymptotic behavior of periodic dynamical systems on Banach space. Ann. Mat. Pura Appl. 86 (1970), 325-330.

[38] Slemrod, M., Asymptotic bahavior of a class of abstract dynamical systems. J. Diff. Eqs. 7 (1970), 584-600.

[39] Slemrod, M., The linear stabilization problem in Hilbert space. J. Funct. Anal. 11 (1972), 334-345.

[40] Slemrod, M., A note on complete controllability and stabilizability of linear control systems in Hilbert space. SIAM J. Control 12 (1974), 500-508.

[41] Slemrod, M., A hereditary partial differential equation with applications in the theory of simple fluids. Arch. Rat. Mech. Anal. 62 (1976), 303-321.

[42] Slemrod, M. & Infante, E. F., An invariance principle for dynamical systems on Banach space. "Instability of Continuous Systems", H. Leipholz ed., pp. 215-221. Springer-Verlag, Berlin 1971.

[43] Walker, J. A., On the application of Liapunov's direct method to linear dynamical systems. J. Math. Anal. Appl. 53 (1976), 187-220.

[44] Walker, J. A., Some results on Liapunov functions and generated dynamical systems. (Preprint).

[45] Walker, J. A. & Infante, E. F., Some results on the precompactness of positive orbits of dynmaical systems. J. Math. Anal. Appl. 51 (1975), 56-67.

[46] Webb, G. F., A bifurcation problem for a nonlinear hyperbolic partial differential equation. (Preprint).

The paper was written while I was visiting the Mathematics Research Center and the Mathematics Department of the University of Wisconsin. I wish to thank my colleagues, too numerous to mention individually, for their hospitality. The research was supported by grants from the following agencies: National Science Foundation MCS 76-07247; Office of Naval Research N00014-76-C-0278; Army Research Office AROD DA-ARO-D-31-124-73-G130 and DAAG 29-75-C0024.

Department of Applied Mathematics
Brown University
Providence, Rhode Island 02912

Results and Open Questions in the Asymptotic Theory of Reaction–Diffusion Equations

Paul C. Fife

A. INTRODUCTION.

This paper is about global bounded solutions of reaction-diffusion equations

$$u_t = D\Delta u + f(u), \tag{1}$$

and more particularly, about these solutions' possible long-time asymptotic states. Here u is an $\mathbb{R}^m$-valued function of $x \in \mathbb{R}^n$ and $t \in \mathbb{R}^+$, D is a matrix with non-negative elements (most commonly diagonal), and Δ is the Laplace operator. Systems of this type have wide application in various physical and biological sciences.

When these systems are used to model natural phenomena, it is often desirable to study the "ultimate" behavior of their solutions. For example, it is important to know which solutions exist globally, which approach constant equilibrium states as $t \to \infty$, and whether other stable spatial and/or temporal structures appear (dissipative structures, in the terminology of Prigogine et al.; see [40], [38], for example). My emphasis will be on discovering conditions under which the latter is the case; and even more specifically, under which x-dependent asymptotic structures appear.

I recently wrote a survey article [24] summarizing a good portion of our knowledge concerning stable x-dependent asymptotic states for reaction-diffusion equations, and the first part of the present paper is a review of this informa-

ISBN 0-12-195250-9

tion (a few new results which have made their appearance since the previous article are also mentioned). The reader wishing more details and references is referred to [24].

Following this, ten problems or suggested directions for further research will be described.

For convenience, I give here a few definitions of terms to follow.

A plane wave solution of (1) is one of the form

$$u(x,t) = U(x \cdot \nu - ct)$$

for a unit vector ν.

Wave trains are plane waves with U periodic; wave fronts are plane waves with U approaching limits $U_{\pm}$ as $z = x \cdot \nu - ct \to \pm\infty$; $U_+ \neq U_-$. Pulses are the same but with $U_+ = U_-$. In either case, it is clear that U_+ and U_- are zeros of f.

Throughout, attention will be restricted to problems without boundary, so that u is defined for all x. Other asymptotic results, some similar to these and some not, have been obtained for problems in bounded domains. Most of the asymptotic structures given here probably approximate solutions of boundary value problems in large domains; this matter generates one of the research problems suggested later in the paper. In small domains, however, no-flux boundary conditions being imposed, Conway, Hoff, and Smoller [13] have shown, again in typical cases, that no x-dependent asymptotic structures exist.

B. RESULTS ON STABLE ASYMPTOTIC STATES.

1. Scalar nonlinear diffusion equations in one space variable

In this case, (1) takes the form

$$u_t = u_{xx} + f(u). \tag{2}$$

Asymptotic states for this equation are defined to be equivalence classes of solutions (defined for all x and large enough t) under the equivalence relation

$$u_1 \sim u_2 \quad \text{if}$$

$$\lim_{t\to\infty} \sup |u_1(x,t) - u_2(\pm x - x_0,\ t - t_0)| = 0$$

for some x_0, t_0, and sign $\pm$.

This is a natural definition because translating (x,t) or changing the sign of x in a solution produces another solution with the same essential properties.

We shall be concerned with C_0-stable asymptotic states (SAS's). This means, given any solution in the class and any sufficiently large t_0, every perturbation of the solution at $t = t_0$ of sufficiently small C_0 norm does not remove the solution from the class. Other types of stability, for example with respect to compact support perturbations, are also important in some contexts [2].

Nonconstant SAS's have been found only in cases when f has three or more zeros, at least two of them "stable" (meaning that f changes sign from positive to negative as u increases through the zero in question). The ones which have been found are wave fronts joining two stable zeros $U_{\pm}$ of f, and certain combinations of such fronts. These combinations can be composed, for example, of two fronts of the same type moving away from each other, or of two or more fronts with different ranges and different speeds. For details, see [25]. One can also envision solutions involving an infinite number of diverging pairs of the first type, spaced at ever increasing intervals. We conjecture that combinations constructed this way account for all possible SAS's of (2).

2. Systems in general

Stable asymptotic states can, of course, be defined in the same way for systems. Some natural conjectures about their totality arise in special cases; but any hope at this stage of classifying them all except in such cases would be quite unreasonable.

Based on the results for the scalar equation, two conjectures immediately come to mind:

(a) If the "kinetic equation" corresponding to (1),

$$\frac{du}{dt} = f(u) \tag{3}$$

has no stable asymptotic state, then neither does (1).

(b) If (3) has two or more stable constant solutions, then a stable wave front for (1) exists.

3. Small amplitude stationary solutions

As we shall see here and later, bifurcation theory can

sometimes be used to construct small amplitude asymptotic states. We suppose that f in (1) depends explicitly on some real parameter λ:

$$u_t = Du_{xx} + f(u,\lambda) \tag{4}$$

and that $f(0,\lambda) \equiv 0$. Thus $u \equiv 0$ is always a solution.

Consider the linearization of (4) about the trivial solution:

$$v_t = Dv_{xx} + f_u(0,\lambda)v. \tag{5}$$

Since it has constant coefficients, its solutions of exponential type are easily found. For some λ_c, we suppose that all decay exponentially in time for $\lambda < \lambda_c$, but that some grow exponentially for $\lambda > \lambda_c$. Then in typical cases, small nontrivial stationary or traveling spatially periodic solutions exist for λ in a one- or two-sided neighborhood of λ_c. I conjecture the following, on the basis of preliminary calculations and by analogy with bifurcation problems in fluid dynamics: in typical cases when the bifurcating solutions are confined to a half neighborhood with $\lambda > \lambda_c$, some are stable, according to a linear stability criterion, and a verifiable test can be set up to specify which they are. In analogous problems from fluid dynamics, various definitions of stability ([30], for example) are used, and we expect that a variety of stability concepts will arise here as well.

4. Large amplitude stationary solutions

A class of equations of the form (1) can be devised, for which larger nonconstant stationary solutions exist and can be proved to be stable in the C_0 sense (this is not possible for scalar equations). A more general class, containing small parameters, exists for which solutions can be constructed by methods of asymptotic analysis, and whose stability can be conjectured on the basis of heuristic arguments. Other than these results and a rather extensive list of computer simulations (many referred to, for example, in [38]), I know of none on stable large amplitude stationary solutions.

5. Wave trains

The existence of small amplitude wave trains can be

proved under certain circumstances similar to those described in section 3 above. Their stability analysis is at the same primitive stage.

A different perturbation method may be used to establish the existence of rapid long wave length wave trains [32], [39] when the kinetic equations (3) have a stable periodic solution. These trains are such that for each fixed x, the solution $u(x,t)$ is a perturbation of the given stable periodic solution of (3). The linear stability of these trains has been analyzed in some cases [32].

Larger amplitude wave trains have also been shown to exist sometimes for two special types of reaction-diffusion equations: those of nerve signal propagation on an axon, and those for predator prey dynamics with spatial migration [12].

6. Plane wave fronts

There are a few categories of systems (1) for which wave fronts connecting two stable zeros of f have been shown to exist, either rigorously or by convincing asymptotic methods. And my conjecture is that they are all stable, though stability has not been proved for any of them.

One such category is that of gradient-like systems, wherein $f = \nabla h$ or $(f, \nabla h) > 0$ for some scalar $h(u)$, with two local maxima [10]. Another is a class of systems with $m = 2$ and small parameters entering in strategic locations [33], [23]. Still other examples come from small amplitude analysis, which shows bifurcation of wave fronts to be possible. Finally, a specific example from population genetics is being analyzed [11].

A number of interesting nonstandard front-like structures have been described by Feinn and Ortoleva [21]. Their analysis is formal and is based on multiple time and length scales, occasioned by the existence of small parameters in their model equations.

7. Pulses

There has been extensive work directed toward establishing the existence and stability of pulse solutions of the various reaction-diffusion systems modeling signal propagation along a nerve axon. As expected, their existence has been easier to establish than their stability, and in fact the stability of pulse solutions for the two simplest such systems has only been proved recently. This was done by Feroe [22] for the McKean equation, and even more recently by McLeod [36] for the FitzHugh-Nagumo system.

8. Other structures

Other stable asymptotic states which have been investigated are target, spiral, and scroll patterns. (See, for example, the recent article by Winfree [46] and its references.) They are the best experimentally documented of all SAS's. Very little analytical work has been performed in behalf of spirals and scrolls. Target patterns have been, and are being, studied for a special class of model reaction-diffusion systems, the $\lambda - \omega$ systems [27], [29], [39]. Though these have simplifying features, the analysis is still very sophisticated. The most difficult case, and the one most in correspondence with observed phenomena, is that when $\lambda' < 0$, $\omega' < 0$. Howard and Kopell [29] have proved the existence of a discrete set of expanding target patterns in this case when certain additional hypotheses of a qualitative nature are satisfied by the functions λ and ω.

C. PROBLEMS.

Current research on reaction-diffusion equations is widespread and varied in direction. I shall not mention all the topics of current interest, but wish to emphasize a few which are in line with the preceding material. More importantly, I shall suggest some significant lines of investigation not presently being pursued.

1. Basic theory for scalar equations

The asymptotic theory of scalar equations (2) is in a fairly satisfactory stage of completeness, in that many important questions have been answered. Nevertheless, basic unanswered ones remain, such as those referred to as conjectures in section B.1:

If f has no stable zeros, is it true that (2) has no stable asymptotic states?

More generally, are all the SAS's of (2) accounted for by wave fronts and combinations of them, in the sense described?

The study of scalar equations enjoys significant advantages over corresponding attempts at studying systems, because (i) comparison techniques, (ii) Lyapunov functional techniques, and (iii) the spectral theory of second order differential operators are available. The application of these tools in the asymptotic theory of (2) is clearly apparent in many papers, including [2], [5], [25], [42], [43]. Lyapunov methods, ordinarily applicable to problems on finite spatial domains, may sometimes be extended to whole-line problems, as was shown in [25].

These advantages suggest that the asymptotic theory of scalar equations in one space variable (at least) may be brought close to completion in the near future.

Wave fronts, clearly shown to play a fundamental role in the asymptotic theory in one space dimension, no doubt perform the same role in higher dimensions, as long as we allow them to be nonplanar. For example, Aronson and Weinberger [3] have demonstrated a type of asymptotic speed of propagation of the effect of a compact support disturbance on a uniform rest state. This strongly suggests a type of takeover process by an expanding closed wavefront. Once the situation with one space variable is understood, one's intuition will readily supply conjectures for higher dimensional problems, and proofs for them should be devised.

2. Aspects of basic theory for systems

The first of the conjectures for scalar equations listed in the preceding section has its analog for systems, in the form B.2 (a). This, and the companion conjecture (b), will probably not be answered definitively for a long time; but results along these lines for special cases may well be reasonable research goals.

3. Existence of large amplitude traveling waves

Traveling plane waves, including trains, fronts, and pulses, are functions of a single variable $z = x \cdot \nu - ct$, and so existence questions for these solutions reduce to existence questions for solutions of certain ordinary differential problems. There has been a considerable amount of success in applying the qualitative theory of ordinary differential equations to these questions in particular cases of especial interest [4], [8], [10], [28] and progress will of course continue as these methods are refined.

4. Stability of large amplitude asymptotic states

It was brought out before that rigorous stability results for $m > 1$ are few and far between. There are some clues available, however, to researchers searching for possible approaches to this difficult problem.

The first clue is that linearized analysis is sometimes enough. For traveling waves $U(x - ct)$, certain properties of the spectrum of the linearized operator $Lv \equiv Dv_{zz} + cv_z + f'(U(z))v$ are sufficient to establish stability in various weighted L_∞ spaces. This result was obtained by Evans ([19], [20]) for nerve conduction equations, and by Sattinger [41] in more generality. The principal required property is that 0 be an isolated simple point in the spectrum, and that the rest of it lie in the negative half plane. These properties may well be true for most or all wave fronts if the weight is chosen correctly (see [42], [43] for applications to scalar equations), and may be usually true in unweighted L_∞ for fronts joining two stable critical points of the kinetic equations. With trains, however, even in the scalar case the spectrum has no isolated points; so for systems it is questionable whether this particular criterion can be used to establish the stability of wave trains.

The second clue is that comparison techniques, which have probably supplied the most useful approach to stability for scalar equations (2), have been partially extended to reaction-diffusion systems. For such results, see [1], [7], [14], [44]. These techniques have been used to establish the stability of uniform rest states in some cases [15], [16], but have not as yet been used to establish the stability of dissipative structures for systems. A worthwhile goal is to refine these methods and fully investigate their applicability to stability of large amplitude asymptotic states.

A final clue is that for one space dimension, comparison methods can sometimes be utilized directly in the phase space to gain some information about the dynamical behavior of the x-trajectories of the solutions in phase space as a function of t. Chueh [6] used this to obtain stability results for some scalar equations; it may be possible for systems as well.

5. Bifurcating structures

We have seen that several types of small amplitude solutions may appear through bifurcation in parameter-dependent problems, when the parameter is near certain critical points. Specifically, the existence of small stationary, wave train, wave front, and pulse solutions has been shown under various circumstances. But many other structures no doubt appear this way. In particular, small-amplitude target and spiral patterns should be looked for, (see [39] for some results here).

Two-dimensional bifurcating patterns have been studied extensively in convective flow problems, and no doubt occur in reaction-diffusion problems as well, though to my knowledge no work has been done here.

The existence of nonperiodic stationary solutions of certain model bifurcation problems has been conjectured by Kirchgässner and Scheurle [31] who also provided a formal asymptotic analysis for them. Analogous questions can be formulated in the present context.

6. Nonhomogeneous systems

In some applications, it is desirable to have information about stable solutions of reaction-diffusion equations in an "inhomogeneous medium." By this, I mean systems (1) in which f and D may depend on x and/or t. A fruitful ground for research is the extension of results about SAS's for the homogeneous case (1) to heterogeneous cases. Of course, some symmetry is then lost and traveling wave solutions as we have defined them do not exist. Therefore one should determine what sort of structures, if any, replace them.

For scalar equations, this analysis has so far been restricted to the study of nonconstant stationary solutions (called clines in population genetical contexts). The existence of clines on the infinite line has been obtained by Conley [9] and Fife and Peletier [26], and their uniqueness and stability by the latter authors. Work on analogs to wave fronts is in progress.

In this latter regard, let us return to the question of what structures replace wave fronts in inhomogeneous media. Suppose, for example, that $f(x,u)$ vanishes for all x, when $u = 0$ or $u = 1$. When f itself is independent of x, a wave front, of course, is a solution $u \in [0,1]$ depending only on the combination $x - ct$ such that (for each t), $u \to 0$ as $x \to -\infty$ and $u \to 1$ as $x \to \infty$. If we replace the requirement that u depend only on $x - ct$ by the requirement that $u(x,t)$ be defined for all $t \in \mathbb{R}$ (negative as well as positive), we obtain a reasonable definition for wave fronts which works in an inhomogeneous medium, as well.

If the characteristic diffusion length $D^{1/2}$ in the scalar equation $u_t = Du_{xx} + f(x,u)$ is small compared to the characteristic length of spatial variation of f (which can be taken as $(\sup|f_x(x,u)|)^{-1}$), then the problem of determining wave fronts can be formally set up as that of determining the terms in an asymptotic expansion. The lowest order term in this expansion is obtained by solving an x-independent wave front problem.

In the case of systems, a general formal approach to wave trains modified by heterogeneities has been given by Ortoleva and Ross [39]. Several questions of rigor could occupy a willing researcher in this connection.

7. Boundary effects

A persistent question in all of the above considerations relates to the relevance of the study of SAS's in all space to phenomena in bounded domains. My conjecture is that in most cases, the solutions described in this paper approximate solutions of boundary value problems for large but bounded domains, the principal needed correction being localized in regions adjacent to the boundary. This has been shown by Howard and Kopell [29] to be true for their expanding target patterns. The analysis of boundary effects, and the related question of interaction of two SAS's, needs to be pursued in detail.

8. Domains of attraction and transient behavior

Along with a knowledge of stable asymptotic states, when they occur, it is of interest to be able to characterize those initial data which evolve into them. A more detailed question is how to describe, possibly through asymptotic analysis, the way a solution approaches its ultimate state. Very little is known about these questions for systems with $m > 1$, and even when $m = 1$, the latter question needs substantial analysis. Work along these lines has been performed by Larson and Lange [34], [35] in connection with Fisher's equation (1) with $f(u) = Cu(1 - u)$, and for certain $\lambda - \omega$ systems.

9. Integrodifferential equations

Integrodifferential and integrodifference equations with many of the properties of reaction-diffusion equations arise as models in such areas as population dynamics, including aspects of epidemiology and genetics, and neurophysics. Mathematical results related to many of those discussed here have been obtained by Weinberger [45], Diekmann [17], Ermentrout and Cowan [18], and others. The study of these problems promises to be a fruitful endeavor, as yet relatively little explored.

10. Nonstandard reaction-diffusion equations

Reaction-diffusion systems (1) sometimes appear as deterministic population dynamic models allowing for spatial migration by means of the diffusion terms. When some stochastic effects, due to the finiteness of the population, are incorporated into the model, its analysis can be expected to be much more difficult.

Nagylaki [37] devised a model which is a pair of nonlinear diffusion equations coupled in a nonstandard manner, and which accounts for some genetic drift (a stochastic effect), as well as deterministic migration and natural selection. This model is for a pair of functions $u(x,t)$, $v(y,z,t)$ (representing a certain expected gene frequency as function of space and time, and the covariance of gene frequency at two locations); it assumes the general form

$$u_t - u_{xx} = f_1(x,u(x,t),v(x,0,t))$$

$$v_t - \frac{1}{2}(v_{yy} + v_{zz}) = f_2(y,z,u(y + z,t),u(y - z,t))v(y,z,t)$$
$$+ \alpha f_3(u(y,t),v(y,0,t))\delta(z),$$

for certain functions f_i (δ is the Dirac delta-function). It reduces to the single deterministic scalar equation (2) (but with f depending on x) if $\alpha = 0$ and $v \equiv 0$.

This model suggests a relevant generalized context in which all the questions we have already posed may again be asked. The generalization consists of allowing the components of u in (1) to be functions of different "space" variables, and coupling them by having f depend on the various components of u evaluated at different arguments. The immediate context in Nagylaki's system is population genetics, so as a first (but difficult!) step, its own clines and wave fronts should be analyzed.

REFERENCES

1. H. Amann (1977), Invariant sets and existence theorems for semilinear parabolic and elliptic systems, J. Math. Anal. Appl.

2. D. G. Aronson and H. F. Weinberger (1975), Nonlinear diffusion in population genetics, combustion and nerve propagation, in: Proceedings of the Tulane Program in Partial Differential Equations and Related Topics, Lecture Notes in Mathematics, No. 446, Springer, Berlin.

3. ________________ (1977), Multidimensional nonlinear diffusion arising in population genetics, Advances in Math., (to appear).

4. G. Carpenter (1977), A geometric approach to singular perturbation problems with applications to nerve impulse equations, J. Differential Equations, 23.

5. N. Chafee and E. F. Infante (1974), A bifurcation problem for a nonlinear partial differential equation of parabolic type, Applicable Anal., 4, 17-37.

6. K. N. Chueh (1975), On the asymptotic behavior of solutions of semilinear parabolic partial differential equations, Ph.D. Thesis, Univ. of Wisconsin.

7. ________, C. Conley, and J. Smoller (1977), Positively invariant regions for systems of nonlinear diffusion equations, Indiana Univ. Math. J., (to appear).

8. C. Conley (1975), On the existence of bounded progressive wave solutions of the Nagumo equation, preprint.

9. ________ (1975), An application of Wazewski's method to a nonlinear boundary value problem which arises in population genetics, Univ. of Wisconsin Math. Research Center Tech. Summary Report 1444.

10. C. Conley (1978), Isolated Invariant Sets and the Morse Index, CBMS/NSF Lecture Notes, SIAM, (to appear).

11. ________ and P. Fife, in preparation.

12. ________ and J. Smoller, in preparation.

13. E. Conway, D. Hoff, and J. Smoller (1977), Large time behavior of solutions of systems of nonlinear reaction-diffusion equations, SIAM J. Appl. Math., (to appear).

14. E. Conway and J. Smoller (1977), A comparison technique for systems of reaction-diffusion equations, Comm. in Partial Differential Equations, 2, 679-697.

15. ________________________ (1977), Diffusion and the predator-prey interaction, SIAM J. Appl. Math., (to appear).

16. ________________________ (1977), Diffusion and the classical ecological interactions: asymptotics, in: Nonlinear Diffusion, Proc. of NSF-CBMS Regional Conference on Nonlinear Diffusion, Research Notes in Math., Pitman, London.

17. O. Diekmann (1977), Threshholds and travelling waves for the geographical spread of infection, preprint.

18. G. B. Ermentrout and J. D. Cowan (1977), Spatial patterns in neural nets, preprint.

19. J. W. Evans (1972), Nerve axon equations: III. Stability of the nerve impulse, Indiana Univ. Math. J., 22, 577-593.

20. __________ (1975), Nerve axon equations: IV. The stable and the unstable impulse, Indiana Univ. Math. J., 24, 1169-1190.

21. D. Feinn and P. Ortoleva (1977), Catastrophe and propagation in chemical reactions, preprint.

22. J. Feroe (1978), Temporal stability of soliton impulse solutions of a nerve equation, Biophys. J. (to appear).

23. P. Fife (1977), Asymptotic analysis of reaction-diffusion wave fronts, Rocky Mountain J. Math., (to appear).

24. _______ (1978), Asymptotic states for equations of reaction and idffusion, Bull. Amer. Math. Soc., (to appear).

25. _______ and J. B. McLeod (1977), The approach of solutions of nonlinear diffusion equations to travelling front solutions, Arch. Rational Mech. Anal., 65, 335-361. Also: Bull. Amer. Math. Soc., 81, (1975), 1075-1078.

26. P. Fife and L. Peletier (1977), Nonlinear diffusion in population genetics, Arch. Rational Mech. Anal., 64, 93-109.

27. J. M. Greenberg (1976), Periodic solutions to reaction-diffusion equations, SIAM J. Appl. Math., 30, 199-205.

28. S. P. Hastings (1976), On travelling wave solutions of the Hodgkin-Huxley equations, Arch. Rational Mech. Anal., 60, 229-257.

29. L. Howard and N. Kopell, in preparation.

30. D. D. Joseph (1976), Stability of Fluid Motions I and II, Springer-Verlag, New York.

31. K. Kirchgässner and J. Scheurle (1976), Periodic and quasiperiodic solutions of $\Delta u + \lambda u + o(u) = 0$, in: Applications of Bifurcation Theory, Academic Press, New York.

32. N. Kopell and L. N. Howard (1973), Plane wave solutions to reaction-diffusion equations, Studies in Appl. Math., 52, 291-328.

33. H. Kurland (1978), Dissertation, University of Wisconsin.

34. D. A. Larson (1978), Transient bounds and time asymptotic behavior of solutions to nonlinear equations of Fisher type, SIAM J. Appl. Math., (to appear).

35. ____________ and C. G. Lange (1977), Transient solutions to some weakly diffusive nonlinear diffusion equations, preprint.

36. J. B. McLeod, in preparation.

37. T. Nagylaki (1978), Random genetic drift in a cline, Proc. Nat. Acad. Sci., (to appear).

38. G. Nicolis and I. Prigogine (1977), Self-organization in Nonequilibrium Systems, Wiley-Interscience, New York.

39. P. Ortoleva and J. Ross (1974), On a variety of wave phenomena in chemical and biochemical oscillations, J. Chem. Phys., 60, 5090-5107.

40. I. Prigogine and G. Nicolis (1967), On symmetry-breaking instabilities in dissipative systems, J. Chem. Phys., 46, 3542-3550.

41. D. Sattinger (1976), On the stability of waves of nonlinear parabolic systems, Advances in Math., 22, 312-355.

42. ____________ (1975), Stability of traveling waves of nonlinear parabolic systems, in: Proc. of VIIth

Inter. Conf. on Nonlinear Oscillations, East Berlin.

43. D. Sattinger (1977), Weighted norms for the stability of traveling waves, J. Differential Equations, (to appear).

44. H. F. Weinberger (1975), Invariant sets for weakly coupled parabolic and elliptic systems, Rend. Mat., 8 (VI), 295-310.

45. ________________ (1977), Asymptotic behavior of a model in population genetics, in: Indiana Univ. Seminar in Applied Math., ed. J. Chadam, Lecture Notes in Mathematics, Springer, Berlin.

46. A. T. Winfree (1977), Stably rotating patterns of reaction and diffusion, preprint.

Supported by National Science Foundation Grant MCS77-02139.

Department of Mathematics and
Program in Applied Mathematics
University of Arizona
Tucson, Arizona 85721

Asymptotic Behavior of Some Evolution Systems

Haim Brezis

Introduction

Let C be a closed convex subset of a Hilbert space H. We denote by $S(t)$ a semi-group of nonlinear contractions on C i.e. $\{S(t)\}_{t>0}$ is a family of mappings from C into itself satisfying: $S(0) = I$, $S(t_1) \circ S(t_2) = S(t_1 + t_2)$, $|S(t)x - S(t)y| \leq |x - y| \; \forall t > 0, \forall x,y \in C$ and $\lim_{t\downarrow 0}|x - S(t)x| = 0 \; \forall x \in C$.

We say that $p \in C$ is an equilibrium point of $S(t)$ provided $S(t)p = p \; \forall t > 0$ and we set

$$F = \{p \in C;\ S(t)p = p \ \forall t > 0\} .$$

It is well known that F is convex; we assume that $F \neq \phi$.

Since, in general, $S(t)x$ does not converge to a limit as $t \to \infty$, it is of interest to consider the behavior as $t \to \infty$ of the ergodic mean $\sigma_t = \frac{1}{t}\int_0^t S(\tau)x \, d\tau$.

In §I we prove that σ_t converges weakly as $t \to \infty$ to a limit σ which can be identified as the unique element in $F \cap \overline{\text{conv}}\ \omega(x)$ (where $\omega(x)$ denotes the weak ω-limit set of $S(t)x$). In addition, σ coincides with the asymptotic center of $S(t)x$ in the sense of Edelstein. In general σ_t does not converge strongly to σ; yet in some special cases of importance, strong convergence holds.

ISBN 0-12-195250-9

In §II we consider gradient flows; i.e. $S(t)x$ denotes the solution of

$$\begin{cases} \frac{du}{dt} + \partial\varphi(u) \ni 0 & \text{a.e. on } (0,\infty) \\ u(0) = x \end{cases}$$

where φ is a convex ℓ.s.c. function on H. As we shall see $S(t)x$ converges weakly as $t \to \infty$ to a limit σ such that $\varphi(\sigma) = \text{Min}\ \varphi$. However, strong convergence does not hold in general; this means that the method of steepest descent applied to a (smooth) convex function converges weakly - but not strongly in general.

In §III we consider the asymptotic behavior of the solution of an equation of the form

$$\frac{du}{dt} + \partial\varphi(u) \ni f(t) \text{ a.e. on } (0,\infty),\ u(0) = x$$

where $f(t)$ is a T-periodic forcing term. Here, the system "converges" as $t \to \infty$ to a T-periodic motion.

In §IV we conclude with a simple application.

Many of the results we discuss are due to J. B. Baillon and to R. Bruck; we also refer to the works of Dafermos-Slemrod [17], Brezis-Browder [12], Pazy [19], [20], [21], Reich [22].

I. Convergence of the ergodic mean.

We begin with a simple result

Theorem 1 ([8]) σ_t converges weakly as $t \to \infty$ to some $\sigma \in F$ - which depends on x.

Remark 1. In general σ_t does not converge strongly (see §II).

In order to identify the limit σ we need some definitions:

a) $\omega(x)$ denotes the weak ω-limit set of $S(t)x$ i.e. $y \in \omega(x)$ iff there is a sequence $t_n \to \infty$ such that $S(t_n)x$ converges weakly to y.

b) The asymptotic center of a bounded function $u(t):[0,\infty) \to H$ is defined as follows. Given $y \in H$, set

$G(y) = \limsup_{t\to\infty} |u(t) - y|^2$ so that G is strictly convex, continuous, $G(y) \to \infty$ as $|y| \to \infty$. Therefore G achieves its minimum on H at a unique point: $AC\{u(t)\}$, called the asymptotic center - a concept introduced by Edelstein.

<u>Theorem 2</u> ([15]). Let $\sigma = \text{weak} \lim_{t\to\infty} \sigma_t$. Then

(1) $\sigma = F \cap \overline{\text{conv}}\ \omega(x)$

(2) $\sigma = AC\{S(t)x\}$.

<u>Remark 2</u>. Instead of σ_t consider now a more general averaging process:

$$\sigma_n = \int_0^\infty S(\tau)x\ a_n(\tau)d\tau$$

where $a_n \in L^1(0,\infty)$, $a_n \geq 0$ and $\int_0^\infty a_n(\tau)d\tau = 1$. Assume that the functions a_n are of bounded variation and $\int_0^\infty |da_n| \to 0$. Then σ_n converges weakly to σ as $n \to \infty$, where σ is the same as in Theorem 1 (σ is independent of a_n).

We begin with a simple Lemma

<u>Lemma 1</u>. Assume $p_1, p_2 \in F$, $q_1, q_2 \in \omega(x)$. Then $(p_1 - p_2, q_1 - q_2) = 0$. In particular $F \cap \overline{\text{conv}}\ \omega(x)$ contains at most one point.

<u>Proof of Lemma 1</u>

Set $u(t) = S(t)x$. Given $p \in F$, the function $|u(t) - p|^2$ is nonincreasing and thus converges as $t \to \infty$ to a limit, say $\ell(p)$.

We have

$$|u(t) - p_1|^2$$

$$= |u(t) - p_2|^2 + 2(u(t) - p_2, p_2 - p_1) + |p_2 - p_1|^2 .$$

Choosing $t = t_n$ such that $u(t_n) \rightharpoonup q_1$ we find

(3) $\ell(p_1) = \ell(p_2) + 2(q_1 - p_2, p_2 - p_1) + |p_2 - p_1|^2$.

Similarly

(4) $\ell(p_1) = \ell(p_2) + 2(q_2 - p_2, p_2 - p_1) + |p_2 - p_1|^2$.

Comparing (3) and (4) leads to the conclusion.

Proofs of Theorem 1, 2 and Remark 2

Let a_n and σ_n be as in Remark 2. Since $\|a_n\|_{L^\infty} \leq \int_0^\infty |da_n| \to 0$ it follows that every weak limit point of σ_n lies in $\overline{\text{conv}}\ \omega(x)$.

We recall that $S(t)$ has a generator in the sense of Komura, Kato, Crandall-Pazy (see e.g. [11]); that is, there exists a maximal monotone operator A with $\overline{D(A)} = C$ such that when $x \in D(A)$, $S(t)x$ coincides with the unique solution $u(t)$ of $\frac{du}{dt} + Au \ni 0$, a.e. on $(0,\infty)$, $u(0) = x$. Assume first $x \in D(A)$ and thus

$$(Av + \frac{du}{dt}(t),\ v - u(t)) \geq 0 \quad \forall v \in D(A)$$

or

$$(Av,\ v - u(t)) \geq \frac{1}{2}\frac{d}{dt}|u(t) - v|^2 .$$

Multiplying by $a_n(t)$ and integrating we find

$$(Av,\ v - \sigma_n) \geq -\frac{1}{2}a_n(0)|x - v|^2 - \frac{1}{2}\int_0^\infty |u(t) - v|^2 da_n .$$

Hence $\forall v \in D(A)$

$$(Av,\ v - \sigma_n) \geq -\|u - v\|^2_{L^\infty} \int_0^\infty |da_n| . \tag{5}$$

By a density argument, (5) holds now even when $x \in \overline{D(A)} = C$.

It follows from (5) that every weak limit point of σ_n lies in F. Indeed let $\sigma_{n_k} \rightharpoonup \sigma$; we have

$$(Av,\ v - \sigma) \geq 0 \quad \forall v \in D(A)$$

and so $0 \in A\sigma$ i.e. $\sigma \in F$.

Consequently every weak limit point of σ_n lies in $F \cap \overline{\text{conv}}\ \omega(x)$ - which consists of a single element.

We prove now that σ coincides with $AC\{S(t)x\}$. Given $y \in H$ we have

$$|u(t) - \sigma|^2 = |u(t) - y|^2 + 2(u(t) - y,\ y - \sigma) + |y - \sigma|^2 .$$

Thus

$$\int_0^\infty |u(t) - \sigma|^2 a_n(t)dt = \int_0^\infty |u(t) - y|^2 a_n(t)dt +$$

$$+ 2(\sigma_n - y,\ y - \sigma) + |y - \sigma|^2 .$$

Since $\sigma \in F$, $|u(t) - \sigma|^2$ converges to a limit as $t \to \infty$, which is simply $G(\sigma)$. Since on the other hand $G(y) = \limsup_{t\to\infty} |u(t) - y|^2$ we find

$$G(\sigma) \leq G(y) - |y - \sigma|^2 \quad \forall y \in H \quad .$$

Consequently $\sigma = AC\{S(t)x\}$.

Remark 3. Let T be a contraction in H having at least one fixed point. Baillon [2] has proved the following (see also [19]):

Theorem 3. The Cesaro mean

$$\sigma_n = \frac{1}{n}(x + Tx + \ldots T^{n-1}x)$$

converges weakly as $n \to \infty$ to a fixed point of T.

Theorem 3 can be viewed as a nonlinear version of the classical ergodic theory of von Neumann, Kakutani, Yosida. Theorem 1 can be derived from Theorem 3 by using a device due to Konishi (see [6]). In general σ_n does not converge strongly.

Theorem 3 has been extended by Baillon to L^p spaces, $1 < p < \infty$, in [6] and subsequently by Bruck [16] to more general spaces. The proofs are very tricky. In view of Konishi's device - which is valid in general Banach spaces - Theorem 1 holds true in L^p spaces $1 < p < \infty$.

Strong convergence

In some special cases σ_t converges strongly as $t \to \infty$. For example when the orbit $\bigcup_{t>0} S(t)x$ is relatively compact - this is the setting of Dafermos-Slemrod [17].

Another surprising condition which implies strong convergence of σ_t is the oddness of $S(t)$ - the fact that oddness has an impact on strong convergence was first observed by Bruck [14] in the case of gradient flows.

Theorem 4 ([4]). Assume C is symmetric and $S(t)$ is odd i.e. $S(t)(-u) = -S(t)u \ \forall u \in C, \forall t > 0$. Then σ_t converges strongly as $t \to \infty$. More generally $\sigma_n = \int_0^\infty S(t)x\, a_n(t)dt$ converges strongly as $n \to \infty$ provided a_n satisfies the assumptions in Remark 2.

Remark 4. It is not known whether Theorem 4 holds in L^p spaces $1 < p < \infty$.

Remark 5. Assume T is an odd contraction on H. Then Baillon [3] has proved that

$$\sigma_n = \frac{1}{n}(x + Tx + \ldots T^{n-1}x)$$

converges strongly to a fixed point of T. Such a result, combined with Konishi's device could be used to prove Theorem 4.

A crucial ingredient in the proof of Theorem 4 is the following:

Lemma 2 ([15], [12], [22]). Let $u(t)$ be a function defined on $(0,\infty)$ with values into H such that: for all $h \geq 0$, $(u(t), u(t+h))$ converges as $t \to \infty$ to a limit, say $\ell(h)$, uniformly in h. Then $\sigma_n = \int_0^\infty u(\tau)\, a_n(\tau)d\tau$ converges strongly as $n \to \infty$ to $AC\{u(t)\}$ provided a_n satisfies the conditions of Remark 2.

For the proof of Lemma 2 we refer to [15], [12] and [22]. Note that a special case of Lemma 2 asserts that if $\{x_n\}$ is a sequence such that (x_n, x_{n+i}) converges as $n \to \infty$ to $\ell(i)$, uniformly in i, then $\sigma_n = \frac{1}{n}(x_0 + x_1 + \ldots x_{n-1})$ converges strongly.

Proof of Theorem 4. Let $t \geq s$ and $x, y \in C$. We have

$$|S(t+h)x - S(t)y| \leq |S(s+h)x - S(s)y|$$

and so

$$|S(t+h)x|^2 - 2(S(t+h)x, S(t)y) + |S(t)y|^2 \leq$$
$$|S(s+h)x|^2 - 2(S(s+h)x, S(s)y) + |S(s)y|^2 .$$

Hence we find

$$2[(S(s+h)x, S(s)y) - (S(t+h)x, S(t)y)]$$
$$\leq |S(s+h)x|^2 + |S(s)y|^2 - |S(t+h)x|^2 - |S(t)y|^2 .$$

Choosing $y = \pm x$ and using the oddness of $S(t)$ leads to

$$2|(S(s+h)x, S(s)x) - (S(t+h)x, S(t)x|$$
$$\leq 2|S(s)x|^2 - 2|S(t+h)x|^2 .$$

But $|S(t)x|$ converges as $t \to \infty$. It follows that $(S(t)x, S(t+h)x)$ is Cauchy as $t \to \infty$, uniformly in h. Therefore Lemma 2 can be applied.

II. Gradient flows

For the class of gradient flows corresponding to convex functions it is not necessary to consider averages of orbits: the orbit itself converges as $t \to \infty$. More precisely let φ be a ℓ.s.c. convex function on H such that $\text{Min}\,\varphi$ is achieved. Set $F = \{p;\ \varphi(p) = \text{Min}\,\varphi\}$ and set $A = \partial\varphi$. We denote by $S(t)x$ the semigroup generated by $-A$ i.e. $S(t)x = u(t)$ is the unique solution of

$$\frac{du}{dt} + Au \ni 0 \quad \text{a.e. on} \quad (0,\infty),\ u(0) = x \quad .$$

Theorem 5 (Bruck [14]). For each $x \in \overline{D(A)}$ $S(t)x$ converges weakly as $t \to \infty$ to some $p \in F$.

Theorem 5 is an obvious consequence of Theorem 1 and the following

Lemma 3. We have for every $x \in \overline{D(A)}$

$$\lim_{t\to\infty} \left|S(t)x - \frac{1}{t}\int_0^t S(\tau)x\,d\tau\right| = 0 \quad . \tag{6}$$

Proof. We have, setting $u(t) = S(t)x$

$$\left|u(t) - \frac{1}{t}\int_0^t u(\tau)\,d\tau\right| = \frac{1}{t}\left|\int_0^t u'(\tau)\ \tau\,d\tau\right| \quad .$$

Therefore it suffices to prove that $\lim_{t\to\infty} t|u'(t)| = 0$. We know (see [10] Theorem 22) that

$$\int_0^\infty \left|\frac{du}{dt}(t)\right|^2 t\ dt \le \frac{1}{2}|\text{dist }(x, F)|^2 \quad .$$

Since the function $t \mapsto \left|\frac{du}{dt}(t)\right|$ is non increasing it follows that

$$\frac{3}{8}\left|\frac{du}{dt}(t)\right|^2 t^2 \le \int_{t/2}^t \left|\frac{du}{dt}(\tau)\right|^2 \tau\ dt \to 0 \quad \text{as} \quad t \to \infty \quad .$$

Remark 6. Baillon [5] has constructed an example of a convex ℓ.s.c. function φ such that $S(t)x$ does not converge strongly as $t \to \infty$. The existence of such an example had been suggested earlier by Kōmura. In fact such a φ can be chosen to be C^1 with grad φ lipschitzian (see [7]). In view of (6), $\sigma_t = \frac{1}{t}\int_0^t S(\tau)x$ does not converge strongly as $t \to \infty$, thus providing a counterexample to the strong convergence of the ergodic mean.

Strong convergence of $S(t)x$ holds in the following cases:

a) For every M the set $\{u \in H, |u| \leq M$ and $\varphi(u) \leq M\}$ is compact (see [10]).

b) The function φ is even (see [14]); here A is odd and we conclude by Theorem 4 and (6) that $S(t)x$ converges strongly.

It would be of interest to find other conditions implying the strong convergence of $S(t)x$. For example, assume φ is even and let $f \in H$ be such that $\underset{z\in H}{\mathrm{Min}}\{\varphi(z) - (f,z)\}$ is achieved.

Let $u(t)$ be the solution of $\frac{du}{dt} + \partial\varphi(u) \ni f, u(0) = x$. Does $u(t)$ converge strongly as $t \to \infty$?

Remark 7. A "discrete" version of Theorem 5 has been proved in [13]: let $A = \partial\varphi$ and let u_n be the sequence defined by the "implicit" scheme

$$\frac{u_{n+1} - u_n}{\tau_n} + Au_{n+1} \ni 0 \qquad \tau_n > 0 \quad .$$

Then $u_n \rightharpoonup p \in F$ provided $\sum_1^\infty \tau_n = \infty$.

III. Periodic forcing

We turn now to the following question. Consider a system governed by a convex potential φ with a periodic forcing term $f(t)$:

$$(7) \qquad \begin{cases} \frac{du}{dt}(t) + \partial\varphi(u(t)) \ni f(t) & \text{a.e. on } (0,\infty) \\ u(0) = u_0 \quad . \end{cases}$$

What happens as $t \to \infty$? The answer is given by the following

Theorem 6 (Baillon-Haraux [9])

Assume $f \in L^2_{\ell oc}(0,\infty; H)$ is T-periodic and assume $u(t)$ remains bounded as $t \to \infty$. Then there exists a T-periodic function $\bar{u}(t)$ satisfying

$$\frac{d\bar{u}}{dt} + \partial\varphi(\bar{u}) \ni f \quad \text{a.e. on} \quad (0,\infty)$$

such that $\underset{t\to\infty}{\text{w-lim}}(u(t) - \bar{u}(t)) = 0$.

In addition

$$\lim_{n\to\infty} \int_0^T \left|\frac{du}{dt}(nT + t) - \frac{d\bar{u}}{dt}(t)\right|^2 dt = 0 \quad .$$

In other words the system evolves "slowly" to a periodic motion. For the proof of Theorem 6 we refer to [9]. The proof involves an interesting Lemma which we describe in a special case:

Lemma 4. Let $f \in L^2(0,T;\, H)$ and let $A = \partial\varphi$. Assume u_n and u are the solutions of

$$\frac{du_n}{dt} + Au_n \ni f \quad \text{a.e. on} \quad (0,T),\ u_n(0) = u_{0n}$$

$$\frac{du}{dt} + Au \ni f \quad \text{a.e. on} \quad (0,T),\ u(0) = u_0$$

with $u_{0n} \to u_0$. Then $u_n \to u$ in $C([0,T];\, H)$ and

$$\int_0^T \left|\frac{du_n}{dt} - \frac{du}{dt}\right|^2 t\, dt \to 0 \ .$$

Proof. We know (see [10]) that $u_n \to u$ in $C([0,T];\, H)$, $\varphi(u_n),\ \varphi(u) \in L^1(0,T)$ and $\sqrt{t}\,\frac{du_n}{dt},\ \sqrt{t}\,\frac{du}{dt} \in L^2(0,T;\, H)$. We have

$$(8) \quad \varphi(u(t)) - \varphi(u_n(t)) \geq \left(f(t) - \frac{du_n(t)}{dt},\ u(t) - u_n(t)\right)$$

and for fixed $v \in D(\varphi)$

$$(9) \quad \varphi(v) - \varphi(u_n(t)) \geq \left(f(t) - \frac{du_n(t)}{dt},\ v - u_n(t)\right).$$

From (9) we deduce

$$(10) \quad \int_0^\varepsilon \varphi(u_n)dt \leq \varepsilon\varphi(v) - \int_0^\varepsilon (f,\ v - u_n)dt - \frac{1}{2}|v - u_n(\varepsilon)|^2 + \frac{1}{2}|v - u_{0n}|^2$$

and from (8)

$$(11) \quad \int_\varepsilon^T \varphi(u_n)dt \le \int_\varepsilon^T \varphi(u)dt + \int_\varepsilon^T (f, u_n - u)dt$$
$$+ \int_\varepsilon^T (\frac{du}{dt}, u - u_n)dt - \frac{1}{2}|u_n(T) - u(T)|^2 + \frac{1}{2}|u_n(\varepsilon) - u(\varepsilon)|^2 .$$

Adding (10) and (11) and using the fact that $u_n \to u$ in $C([0,T]; H)$ we find

$$\limsup_{n\to\infty} \int_0^T \varphi(u_n)dt \le \varepsilon\varphi(v) - \int_0^\varepsilon (f, v - u)dt - \frac{1}{2}|v - u(\varepsilon)|^2$$
$$+ \frac{1}{2}|v - u_0|^2 + \int_\varepsilon^T \varphi(u)dt \quad .$$

As $\varepsilon \to 0$ we obtain

$$(12) \quad \limsup_{n\to\infty} \int_0^T \varphi(u_n)dt \le \int_0^T \varphi(u)dt \quad .$$

On the other hand we have

$$(13) \quad \int_0^T |\frac{du_n}{dt}|^2 t\, dt + T\varphi(u_n(T)) - \int_0^T \varphi(u_n)dt = \int_0^T (f, \frac{du_n}{dt})t\, dt$$

$$(14) \quad \int_0^T |\frac{du}{dt}|^2 t\, dt + T\varphi(u(T)) - \int_0^T \varphi(u)dt = \int_0^T (f, \frac{du}{dt})t\, dt \quad .$$

Combining (12, (13), (14) and the fact that $\sqrt{t}\,\frac{du_n}{dt}$ converges weakly in $L^2(0, T; H)$ to $\sqrt{t}\,\frac{du}{dt}$ we see that $\limsup_{n\to\infty} \int_0^T |\frac{du_n}{dt}|^2 t\, dt \le \int_0^T |\frac{du}{dt}|^2 t\, dt$, and therefore

$$\lim_{n\to\infty} \int_0^T |\frac{du_n}{dt} - \frac{du}{dt}|^2 t\, dt = 0 \ .$$

Remark 8. Using a similar device, H. Attouch [1] has proved the following. Let $A_n = \partial\varphi_n$ and $A = \partial\varphi$ be such that $A_n \to A$ in the sense of graphs i.e. $(I + A_n)^{-1} \to (I + A)^{-1}$. Assume u_n and u are the solutions of

$$\frac{du_n}{dt} + A_n u_n \ni f \quad \text{a.e. on} \quad (0,T),\ u_n(0) = u_{0n}$$

$$\frac{du}{dt} + Au \ni f \qquad \text{a.e. on} \quad (0,T),\ u(0) = u_{0n}$$

with $u_{0n} \to u_0$.

Then $u_n \to u$ in $C([0,T]; H)$ (this is just the nonlinear version of the Trotter-Kato theorem) and also

$$\int_0^T \left|\frac{du_n}{dt} - \frac{du}{dt}\right|^2 t\, dt \to 0.$$

Remark 9. Let $f \in L^2_{\ell oc}(0,\infty ; H)$, let $A = \partial\varphi$ and let u, v be the solutions of

$$\frac{du}{dt} + Au \ni f \quad \text{a.e. on} \quad (0,\infty),\ u(0) = u_0$$

$$\frac{dv}{dt} + Av \ni f \quad \text{a.e. on} \quad (0,\infty),\ v(0) = v_0 \quad .$$

Does $u(t) - v(t)$ converge weakly to a limit as $t \to \infty$? Note that the answer is positive in case f is T-periodic and $u(t)$ remains bounded. Indeed there are two T-periodic solutions $\bar{u}(t)$ and $\bar{v}(t)$ such that $\operatorname{w-lim}_{t\to\infty}(u(t) - \bar{u}(t)) = 0$, $\operatorname{w-lim}_{t\to\infty}(v(t) - \bar{v}(t)) = 0$. Since the difference of two T-periodic solutions is constant (see [9]) it follows that $\operatorname{w-lim}_{t\to\infty}(u(t) - v(t))$ exists.

IV. An example

Let Ω be a smooth bounded domain in $\mathbb{R}^N$. Given $\psi(x) \in L^2(\Omega)$ with $\psi \geq 0$ on Ω and $u_0 \in H_0^1(\Omega)$, find a function $u(x,t)$ satisfying

$$(15) \qquad \begin{cases} u_t = \min\{\Delta u, \psi\} & \text{on} \quad \Omega \times (0,\infty) \\ u = 0 & \text{on} \quad \partial\Omega \times (0,\infty) \\ u(x,0) = u_0(x) & \text{on} \quad \Omega \end{cases}$$

(related questions occur in heat control, see [18] Chap. 2).

Theorem 7. There exists a unique solution of (15). In addition $u(x,t)$ converges weakly in H_0^1, as $t \to \infty$, to a function $u_\infty(x)$ satisfying

$$(16) \qquad \min\{\Delta u_\infty, \psi\} = 0 \quad .$$

Remark 10. It is clear that any function $u_\infty \in H_0^1$ satisfying (16) is an equilibrium and that there are many such functions. We don't know how to identify $\lim_{t\to\infty} u(x,t)$ (in terms of u_0) among all the equilibria. Also we don't know whether $u(x,t)$ converges strongly in H_0^1.

Proof. Let $K = \{v \in H_0^1(\Omega),\ v \le \psi \ \text{a.e.}\}$. On $H = H_0^1(\Omega)$ we consider the scalar product

$$a(u,v) = \int_\Omega \operatorname{grad} u \operatorname{grad} v \, dx \quad .$$

Problem (15) can be expressed in a weak form as:

$$u_t \in K$$

$$\int_\Omega u_t(v - u_t)dx + a(u,\ v - u_t) \ge 0 \quad \forall v \in K,\ \forall t > 0$$

or equivalently

$$a(u,\ v - u_t) + \varphi(v) - \varphi(u_t) \ge 0 \quad \forall v \in H_0^1,\ \forall t > 0 \tag{17}$$

where φ is a convex $\ell.s.c.$ function defined on H by

$$\varphi(v) = \begin{cases} \frac{1}{2}\int_\Omega |v|^2 dx & \text{if } v \in K \\ +\infty & \text{if } v \notin K \ . \end{cases}$$

But (17) can be written as $-u \in \partial\varphi(u_t)$ or $u_t - \partial\varphi^*(-u) \ni 0$.

Now we are reduced to the abstract setting and we can use Theorem 5. Note that u_∞ is an equilibrium provided $-u_\infty \in \partial\varphi(0)$ i.e.

$$\frac{1}{2}\int_\Omega v^2 \, dx \ge -\ a(u_\infty,\ v) \qquad \forall v \in K$$

or equivalently $\int_\Omega \operatorname{grad} u_\infty \operatorname{grad} v \, dx \ge 0 \quad \forall v \in K.$

Remark 11. Instead of (15) consider now a problem with two side-constraints on u_t: find $u(x,t)$ satisfying

$$\begin{cases} |u_t| \le \psi & \text{on } \Omega \times (0,\infty) \\ u_t - \Delta u = 0 & \text{on } [|u_t| < \psi] \\ u_t - \Delta u \le 0 & \text{on } [u_t = \psi] \\ u_t - \Delta u \ge 0 & \text{on } [u_t = -\psi] \\ u = 0 & \text{on } \partial\Omega \times (0,\infty) \\ u(x,0) = u_0(x) & \text{on } \Omega \ . \end{cases} \tag{18}$$

The same kind of proof as above shows that there is a unique solution u of (18) and that $u(x,t)$ converges <u>strongly</u> in H_0^1 to a function u_∞ satisfying $\Delta u_\infty = 0$ on $[\psi > 0]$. Here strong convergence holds since φ is even.

REFERENCES

[1] H. Attouch, Convergence de fonctionnelles convexes, to appear.

[2] J. B. Baillon, Un théorème de type ergodique pour les contractions nonlinéaires dans un espace de Hilbert, C. R. Acad. Sc. 280 (1975) p. 1511-1514.

[3] J. B. Baillon, Quelques propriétés de convergence asymptotique pour les contractions impaires, C. R. Acad. Sc. 283 (1976) p. 587-590.

[4] J. B. Baillon, Quelques propriétés de convergence asymptotique pour les semi-groupes de contractions impaires, C. R. Acad. Sc. 283 (1976) p. 75-78.

[5] J. B. Baillon, Un exemple concernant le comportement asymptotique de la solution du problème $\frac{du}{dt} + \partial\varphi(u) \ni 0$, to appear J. Funct. Anal.

[6] J. B. Baillon, Comportement asymptotique des contractions nonlinéaires et des semi-groupes continus de contractions dans les espaces de Banach, C. R. Acad. Sc. (1978) and detailed paper to appear.

[7] J. B. Baillon, Thèse, Université Paris VI (1978).

[8] J. B. Baillon - H. Brezis, Une remarque sur le comportement asymptotique des semi groupes nonlinéaires, Houston J. Math. 2 (1976) p. 5-7.

[9] J. B. Baillon - A. Haraux, Comportement à l'infini pour les équations d'evolution avec forcing periodique, Archive Rat. Mech. Anal. 67 (1977) p. 101-109.

[10] H. Brezis, Monotonicity methods, in Contributions to Nonlinear Functional Analysis, Zarantonello ed. Acad. Press (1971).

[11] H. Brezis, Opérateurs maximaux monotones, Lecture Notes No. 5, North Holland (1973).

[12] H. Brezis - F. Browder, Remarks on nonlinear ergodic theory, Adv. in Math. 25 (1977) p. 165-177.

[13] H. Brezis - P. L. Lions, Produits infinis de résolvantes, Israel J. Math.

[14] R. Bruck, Asymptotic convergence of nonlinear contraction semi groups in Hilbert spaces, J. Funct. Anal. 18 (1975) p. 15-26.

[15] R. Bruck, On the almost convergence of iterates of a nonexpansive mapping in a Hilbert space and the structure of the weak ω-limit set, Israel J. Math. 29 (1978) p. 1-17.

[16] R. Bruck, A simple proof of the mean ergodic theorem for nonlinear contractions in Banach spaces.

[17] C. Dafermos - M. Slemrod, Asymptotic behavior of nonlinear contraction semi groups, J. Funct. Anal. 13 (1973) p. 97-106.

[18] G. Duvaut - J. L. Lions, Les inéquations en mécanique et en physique, Dunod (1972).

[19] A. Pazy, On the asymptotic behavior of iterates of nonexpansive mappings in Hilbert space, Israel J. Math. 26 (1977) p. 197-204.

[20] A. Pazy, On the asymptotic behavior of semigroups of nonlinear contractions in Hilbert space, J. Funct. Anal.

[21] A. Pazy, The asymptotic behavior of semigroups of nonlinear contractions having large sets of fixed points. Proc. Royal Soc. Edinburgh.

[22] S. Reich, Almost convergence and nonlinear ergodic theorems. J. Approx. Theory.

Department of Mathématiques
Université Paris VI
75230 Paris, France

Trotter's Product Formula for Some Nonlinear Semigroups

Tosio Kato

1. Introduction

Trotter's product formula for a pair $\{e^{-tA}\}$, $\{e^{-tB}\}$ of (linear) C_0-semigroups on a Banach space X was first proved by Trotter [1], although the trivial case of semigroups with bounded generators may have been noticed earlier. Formally, the formula asserts that

(1.1) $$[e^{-(t/n)B} e^{-(t/n)A}]^n \to e^{-t(A+B)}, \quad n \to \infty .$$

Trotter proved (1.1) when a certain <u>domain condition</u> for A, B is satisfied. He assumed, essentially, that

(1.2) $\begin{cases} -(A+B) \text{ defined on } D(A) \cap D(B) \text{ is closable} \\ \quad \text{and the closure is the generator of a } C_0\text{-semigroup,} \end{cases}$

although the precise condition he assumed was slightly stronger than (1.2).

This theorem of Trotter is very general in that X is an arbitrary Banach space. It may have been somewhat disappointing to those who wished to have the convergence in (1.1) without knowing beforehand that $-(A+B)$ is a generator.

This situation has not improved much since, although condition (1.2) has been extended in many directions. (For an extensive study of the product formula, see Chernoff [2]). Indeed, in almost all cases in which (1.1) was proved, $-(A+B)$ was known or assumed to be a generator, although the definition of $A+B$ need not be so simple as in (1.2). One of few exceptions I know of is a result of Nelson [3], in which the

ISBN 0-12-195250-9

convergence (1.1) is proved directly for $A = -\Delta$ (negative Laplacian) and $B = V(x)$ (operator of multiplication with a complex-valued, singular function) in $X = L^2(R^m)$, so that the generator $-(A+B)$ can be defined _as the result of_ (1.1). His convergence proof depends on the use of the Wiener integral. (Another proof based on the construction of $A+B$ as a generator was recently given by the author [4].)

Even if the situation mentioned above slightly detracts from the usefulness of (1.1), it is still interesting in its own right and is useful for many purposes, including numerical analysis. It gives a concrete means of computing $\{e^{-t(A+B)}\}$ from $\{e^{-tA}\}$ and $\{e^{-tB}\}$. In this connection, it is useful to generalize (1.1) to

$$[V(t/n)U(t/n)]^n \to e^{-t(A+B)}, \quad n \to \infty, \tag{1.3}$$

where $\{U(t)\}$, $\{V(t)\}$ are certain approximating families to $\{e^{-tA}\}$, $\{e^{-tB}\}$, respectively. A typical example of $U(t)$ is the resolvent $U(t) = (1+tA)^{-1}$. Other examples will be given below in connection with more precise formulations of (1.3).

A standard tool for proving convergence like (1.3) is CHERNOFF'S LEMMA [2,5,6]. _Let_ $\{F(t);\ t>0\}$ _be a family of (linear) nonexpansive operators on_ X _such that for each_ $x \in X$

$$[1 + \lambda t^{-1}(1-F(t))]^{-1}x \to (1 + \lambda C)^{-1}x, \quad t \downarrow 0, \tag{1.4}$$

for some (and, equivalently, all) $\lambda > 0$, _where_ $-C$ _is the generator of a_ C_0_-semigroup on_ X. _Then_

$$F(t/n)^n x \to e^{-tC}x, \quad n \to \infty, \tag{1.5}$$

uniformly in each finite interval $t \in [0,T]$.

With this lemma, the proof of Trotter's theorem stated above is quite easy.

In applying Chernoff's lemma to the proof of (1.3) with $F(t) = V(t)U(t)$, it is essential to find C, which should be equal to $A+B$ in some sense but which need not be equal to the one given in (1.2). No general method is known to find C

even if it exists. In the case when A, B are <u>nonnegative</u> in a certain sense, however, it turns out that one can give rather general results. Here I want to discuss results of this type and their generalization to nonlinear semigroups.

2. Linear Sectorial Generators.

First assume that A, B are nonnegative selfadjoint operators in a Hilbert space $X = H$. Obviously $-A, -B$ generate semigroups of nonexpansive, nonnegative selfadjoint operators. In this case one can define a sum $A\dot{+}B$ (called the <u>form sum</u>) by way of <u>quadratic forms</u> even when $D(A) \cap D(B)$ is small (it can be $\{0\}$). One first defines the quadratic form

$$(2.1) \quad \begin{cases} c'[u] = \|A^{1/2}u\|^2 + \|B^{1/2}u\|^2 \\ \text{for } u \in D(c') = D(A^{1/2}) \cap D(B^{1/2}) . \end{cases}$$

Then there is a unique nonnegative selfadjoint operator $C' = A\dot{+}B$ in the Hilbert space $H' = \text{cl}.D(c')$ such that

$$(2.2) \quad c'[u] = \|C'^{1/2}u\|^2 \quad , \quad u \in D(c') = D(C'^{1/2}) \quad .$$

Thus $A\dot{+}B$ is selfadjoint in H if and only if $D(A^{1/2}) \cap D(B^{1/2})$ is dense in H (which may be true even if $D(A) \cap D(B) = \{0\}$). $A\dot{+}B$ has no meaning in $H'^{\perp}$, but one may imagine that $-(A\dot{+}B) = -\infty$ there so that the semigroup it generates is zero. Indeed, such an interpretation is justified by a more precise result given by

THEOREM I [7,8]. *Let* $U(t) = f(tA)$, $V(t) = g(tB)$, *where* f, g *are real-valued Baire functions on* $[0,+\infty[$ *such that*

$$(2.3) \quad 0 \le f(s) \le 1, \quad f(0) = 1, \quad f'(0) = -1 \quad ,$$

and similarly for g. *Then*

$$(2.4) \quad [V(t/n)U(t/n)]^n \to e^{-tC'}P' \quad , \quad t > 0, \quad n \to \infty \quad ,$$

strongly, where P' *is the orthogonal projection of* H *onto* H'. *The convergence is uniform on any finite interval* $[0,T]$ *if* (2.4) *is applied to* $u \in H'$ (*so that* $P'u = u$).

COROLLARY. *The limit* (2.4) *always exists. It forms a* C_0*-semigroup on* H *if and only if* $D(A^{1/2}) \cap D(B^{1/2})$ *is dense in* H (*so that* $H' = H, P' = 1$).

The possibility that the limit in (2.4) may be zero was noticed by Chernoff [2] by a special example. It is the case if and only if $D(A^{1/2}) \cap D(B^{1/2}) = \{0\}$. Also [2] contains a number of sufficient conditions (some of them due to the author) for (2.4) to be true with $P' = 1$. Other conditions of similar kind were given by Beliy and Semenov [9]. All these conditions are now superseded by Theorem I and its Corollary.

Theorem I can be generalized to the case in which A, B are no longer selfadjoint but are m-sectorial (see [6] for the definition). The proof (given in [8]) is due to B. Simon, and is an ingenious application of analytic continuation. In this case C' is also m-sectorial in H', being associated with a nonreal quadratic form $c' = a+b$ defined on $D(c') = D(a) \cap D(b)$, where a, b are the quadratic forms associated with A, B, respectively. It is important to note that in general $D(a)$ is different from $D(A^{1/2})$ (see McIntosh [10]).

There are certain results in operator theory which have been proved only by means of Trotter's formula. An example is the Segal-Golden-Thompson formula (see e.g. Reed and Simon [11]) for nonnegative selfadjoint operators A, B:

$$\|e^{-(A\dot{+}B)}\|_p \leq \|e^{-A}e^{-B}\|_p \quad , \quad 0 < p \leq \infty \quad , \tag{2.5}$$

where $\| \ \|_p$ for $p < \infty$ is the p-"norm" for compact operators (the p-th root of the sum of the p-th powers of the singular values); for $p = \infty$ it is the usual operator norm. In (2.5) we assume, for simplicity, that $D(A^{1/2}) \cap D(B^{1/2})$ is dense in H so that $A\dot{+}B$ is also selfadjoint in H.

A proof of (2.5) for $p = \infty$ is given in [11, p.261] under the same condition as in (1.2). This condition is used to justify the application of Trotter's formula (1.1). Hence the proof is valid in the more general case stated above. The same remark applies to other values of p.

3. Nonlinear Semigroups.

Since the creation by Kōmura [12] of the theory of semigroups of nonlinear nonexpansive operators, there arose attempts to extend Trotter's formula to such semigroups. The first result in this direction was given by Mermin [13], where a formula of the form (1.3) is proved for $U(t) = (1+tA)^{-1}$, $V(t) = (1+tB)^{-1}$, where A, B are m-accretive operators in a Banach space X with X^* uniformly convex and where B is assumed to be small relative to A in a certain sense. More general results were given by Brezis and Pazy [14,15], where the fundamental generalization of Chernoff's lemma to nonlinear semigroups is also given (see also Brezis [16]). The generalized lemma has almost the same form as in the linear case, with the main difference that (1.4) is required only for x in the closure of $D(C)$ and (1.5) is asserted only for such x. The product formulas (1.3) proved in [14,15] cover the cases when $U(t)$, $V(t)$ are the resolvents (as in [13]) and also when they are semigroups $\{e^{-tA}\}$, $\{e^{-tB}\}$; the domain condition assumed is that $D(A) \cap D(B)$ is sufficiently large, roughly corresponding to (1.2).

Once it is known that Theorem I is true for linear operators A, B without any domain conditions, it is expected that a similar result should hold for nonlinear semigroups. A natural nonlinear generalization of a nonnegative selfadjoint operator in H is the subdifferential $d\phi$ of a lower semicontinuous, proper convex function ϕ on H to $]-\infty,+\infty]$. (We denote by Φ the set of all such functions ϕ.) It turns out that Theorem I extends to the case of semigroups generated by operators of the form $-d\phi$ with simple modifications.

To state the theorem, it is convenient to introduce the notion of a ϕ-<u>family</u>. Let $\phi \in \Phi$. A family $\{U(t);\ t>0\}$ of nonlinear nonexpansive operators on H will be called a ϕ-family if

$$\text{(3.1)} \quad \phi(v) \geq \phi(U(t)y) + t^{-1}(v-y, y-U(t)y) + \gamma(2t)^{-1}\|y-U(t)y\|^2$$

for every $v, y \in H$ and $t > 0$, where $\gamma > 0$ is a constant. The largest possible γ will be called the ϕ-<u>index</u> of the family.

The following are examples of ϕ-families and their ϕ-indices:

$$\text{(3.2)} \quad \begin{aligned} U(t) &= [1+(t/k)d\phi]^{-k}, & \gamma &= 1+k^{-1}, \quad k = 1,2,3,\ldots\ldots \\ U(t) &= e^{-td\phi}P, & \gamma &= 1, \end{aligned}$$

where P denotes the (nonlinear) projection of H onto $\mathrm{cl.}D(\phi)$, where $D(\phi)$, the effective domain of ϕ, is a convex set.

THEOREM II. Let $\phi_j \in \Phi$, $j = 1,2,\ldots,N$ such that $\phi = \phi_1 + \cdots + \phi_N \not\equiv +\infty$. Let $D_j = D(\phi_j)$, $D = D(\phi)$, $E_j = \mathrm{cl}.D_j$, $E = \mathrm{cl}.D$, and let P_j, P be the projections of H onto E_j, E, respectively. Let $\{U_j(t)\}$ be a ϕ_j-family with ϕ_j-index γ_j, $j = 1,2,\ldots,N$. Assume that one of the following conditions is satisfied.

(i) $\gamma_j \geq 1$ for all $j = 1,\ldots,N$, and for each k with $\gamma_k = 1$ we have

$$U_k(t)u = U_k(t)P_k u, \quad u \in H \quad . \tag{3.3}$$

(ii) There is a k such that $\gamma_j > 1$ for all $j \neq k$, and

$$(\gamma_k - 1) \sum_{j \neq k} (\gamma_j - 1)^{-1} > -1 \quad . \tag{3.4}$$

Then

$$\lim_{n\to\infty}[U_N(t/n)\cdots U_1(t/n)]^n x = e^{-td\phi}x, \quad t \geq 0, \; x \in E \quad , \tag{3.5}$$

the convergence being uniform in $t \in [0,T]$ for any $T < \infty$.

REMARK. If $N = 2$, (ii) is equivalent to $\gamma_1 + \gamma_2 > 2$; (3.3) is required only when $\gamma_1 = \gamma_2 = 1$; none of (i), (ii) is true if $\gamma_1 + \gamma_2 < 2$. If $N = 1$, disregard (i), (ii) and simply assume $\gamma_1 > 0$.

Without going into the proof of Theorem II, which is given in Kato and Masuda [17], I shall restrict myself to giving some remarks.

(a) The proof is easier in the case (ii).

(b) Unlike Theorem I, Theorem II asserts nothing about (3.5) when x is not in $E = \mathrm{cl}.D$. A simple generalization by admitting all $x \in H$ and replacing x by Px on the right does not work (see [17]).

(c) Except for remark (b), Theorem II essentially contains Theorem I as a special case (in particular when U, V are semigroups). A great advantage of Theorem II is that $N \geq 3$ is permitted, whereas the proof given in [8] for Theorem I does not seem to generalize easily to the case $N \geq 3$.

(d) Looked at more closely, however, Theorems I, II are not comparable because there are differences in the assumptions on the U_j — condition (2.3) on the one hand and condition (3.1) on the other. For example, consider the family

$$(3.6) \quad U(t) = f(tA), \quad f(s) = [1 - (s/2)][1 + (s/2)]^{-1} ,$$

which appears in the so-called Crank-Nicholson scheme.* $\{U(t)\}$ is not a $\emptyset$-family for $A = d\emptyset$ so it cannot be admitted in Theorem II. It does not satisfy condition (2.3) either. But an inspection of the proof of Theorem I shows that it is admissible in the linear case provided the other family $\{V(t)\}$ satisfies (2.3). Thus it is desirable to generalize Theorem II to admit such a family.

REFERENCES

1. H. F. Trotter, On the product of semi-groups of operators, Proc. Amer. Math. Soc. 10 (1959), 545-551.
2. P. R. Chernoff, Product formulas, nonlinear semigroups, and addition of unbounded operators, Mem. Amer. Math. Soc. 140, 1974.
3. E. Nelson, Feynman integrals and the Schrödinger equation, J. Mathematical Phys. 5 (1964), 332-343.
4. T. Kato, On some Schrödinger operators with a singular complex potential, Ann. Scuola Normale Sup., to appear.
5. P. R. Chernoff, Note on product formulas for operator semi-groups, J. Functional Anal. 2 (1968), 238-242.
6. T. Kato, Perturbation theory for linear operators, Springer 1966, 1976.
7. T. Kato, On the Trotter-Lie product formula, Proc. Japan Acad. 50 (1974), 694-698.
8. T. Kato, Trotter's product formula for an arbitrary pair of selfadjoint contraction semigroups, Advances in Math., to appear.
9. A. G. Beliy and Y. A. Semenov, A criterion for the convergence of semigroup product (preprint in Russian), Kiev 1974.

* In the nonlinear case this means $U(t) = [1-(t/2)A][1+(t/2)A]^{-1}$, which is nonexpansive for $t > 0$ if A is m-accretive in a Hilbert space.

10. A. McIntosh, On the comparability of $A^{1/2}$ and $A^{*1/2}$, Proc. Amer. Math. Soc. 32 (1972), 430-434.

11. M. Reed and B. Simon, Methods of modern mathematical physics, Volume II, Academic Press 1975.

12. Y. Kōmura, Nonlinear semi-groups in Hilbert space, J. Math. Soc. Japan 19 (1967), 493-507.

13. J. Mermin, Accretive operators and nonlinear semigroups in spaces with uniformly convex dual, Thesis, University of California, Berkeley, 1968.

14. H. Brezis and A. Pazy, Semigroups of nonlinear contractions on convex sets, J. Functional Anal. 6 (1970), 237-281.

15. H. Brezis and A. Pazy, Convergence and approximation of semigroups of nonlinear operators in Banach spaces, J. Functional Anal. 9 (1972), 63-74.

16. H. Brezis, Opérateur maximaux monotones et semi-groupes de contractions dans les espaces de Hilbert, North-Holland 1973.

17. T. Kato and K. Masuda, Trotter's product formula for nonlinear semigroups generated by the subdifferentials of convex functionals, J. Math. Soc. Japan, to appear.

Supported in part by NSF Grant MCS 76-04655.

Department of Mathematics
University of California-Berkeley
Berkeley, California 94720

Application of Nonlinear Semigroup Theory to Certain Partial Differential Equations

Lawrence C. Evans

Introduction

This paper comprises, first of all, a survey of the main facts about nonlinear semigroups in arbitrary Banach spaces and, secondly, an exposition of some applications and extensions of this general theory to certain nonlinear partial differential equations of parabolic type. Our objective here is to discuss the scope of applicability of the abstract considerations to several concrete problems, to explain what useful information semigroup theory provides and what it does not, and to demonstrate how in certain cases various p.d.e. techniques can be used to extend our theoretical understanding. We will especially emphasize this last topic, the interplay between abstract and *ad hoc* methods.

Section 1 contains the basic working information about nonlinear semigroups. Here we discuss the Crandall-Liggett generation theorem, various regularity and perturbation results, the nonlinear Chernoff theorem, and several related topics, all selected with a view towards the applications presented in Section 2. These are (a) the porous medium and related equations, (b) certain variational and quasi-variational inequalities of evolution, and (c) Bellman's equation of dynamic programming. In reporting on these subjects we will

ISBN 0-12-195250-9

discuss the past and ongoing work of several people, including H. Brezis, M. Crandall, A. Pazy, the author, and especially Ph. Benilan. Our style will be heuristic, entailing fairly detailed explanations at certain points and considerable leaps of faith at others. To simplify the discussion we have stated almost every theorem under stronger hypotheses than are really required.

1. Nonlinear Semigroup Theory

All the mathematics discussed in this section was developed owing to interest in the seemingly quite simple initial value problem

$$\text{(IVP)} \quad \begin{cases} (1.1) \quad \frac{du}{dt}(t) + A(u(t)) = f(t) \quad 0 < t < \infty \\ (1.2) \quad u(0) = x_0. \end{cases}$$

To make sense of this differential equation let us assume that we are given a real Banach space X (with norm denoted by $\| \ \|$), an operator A mapping some domain $D(A) \subseteq X$ back into X, an element $x_0 \in \overline{D(A)}$, and a function $f\colon [0,\infty) \to X$. The first problem is to find a function $u\colon [0,\infty) \to X$ solving (IVP) in some appropriate sense. Having found such a solution, we then of course ask about uniqueness, regularity, behavior with respect to perturbations of x_0, A, and f, etc., etc.

In applications to partial differential equations the operator A is typically nonlinear, not everywhere defined, discontinuous (involving derivatives in variables other than t), and possibly even multivalued, whereas X may not be a Hilbert space or even be reflexive. In such situations the questions posed above regarding the solvability, uniqueness, and so forth are quite difficult, and in general have no satisfactory answers.

It is therefore remarkable that for a certain class of nonlinear operators A the fortunate situation prevails that (1) (IVP) does indeed have a unique solution u (although existing perhaps only in some generalized sense), and (2) many interesting nonlinear partial differential equations can be realized in the form (1.1), (1.2) for such an operator A. This collection comprises the m-accretive mappings on X:

Definition. An operator $A: D(A) \subset X \to X$ is called accretive if

$$(1.3) \qquad \|x - \hat{x}\| \leq \|x - \hat{x} + \lambda(A(x) - A(\hat{x}))\|$$

for all $x, \hat{x} \in D(A)$, $\lambda > 0$. If, in addition,

(1.4) Range $(I + \lambda A) = X$ for some (equivalently, for all) $\lambda > 0$,

then A is m-accretive. (A is also said to be, respectively, monotone and maximal monotone in the case X is a Hilbert space.)

As a first and simple example the reader should verify that

(1.5) $A \equiv I + T$ is m-accretive, whenever T is an everywhere defined contraction on X.

More interesting examples will be provided later.

The key idea for solving the initial value problem (1.1), (1.2) when A is m-accretive is to approximate (IVP) by a sequence of problems $(IVP)_n$ in which the t-derivative is replaced by a difference quotient of given step size $\lambda_n > 0$ and the inhomogeneous term $f(t)$ is replaced by a step function approximation $f^n(t)$ ($= f_k^n$ for $k\lambda_n \leq t < (k+1)\lambda_n$):

$$(IVP)_n \qquad \begin{cases} \dfrac{x_k^n - x_{k-1}^n}{\lambda_n} + A(x_k^n) = f_k^n & k = 1,2,\ldots \\ x_0^n = x_0. \end{cases}$$

Since A satisfies (1.4) we can solve recursively for the x_k^n ($k = 1,2,\ldots$) and thereby construct the piecewise constant approximate solution $u^n(t)$ ($= x_k^n$ for $k\lambda_n \leq t < (k+1)\lambda_n$). The next theorem is the fundamental assertions that the approximations converge:

Generation Theorem. Let A be an m-accretive operator and $x_0 \in \overline{D(A)}$. Suppose that $f \in L^1(0,T; X)$ for some $0 < T < \infty$, and that the approximations $f^n \to f$ in $L^1(0,T;\, X)$ as $n \to \infty$, $\lambda_n \to 0$.

Then the $u^n(t)$ converge uniformly on $[0,T]$ to a limit function $u(t)$.

This result is proved in [22]; the basic ideas originate with the proof of Crandall and Liggett [24] for the case $f \equiv f^n \equiv 0$. When $f \equiv 0$, we sometimes write $u(t) = S(t)x_0$ (to display explicitly the dependence on t and x_0) and call the family of operators $S(t)$ $(t \geq 0)$ so defined the semigroup generated by A.

It is worth stressing here that no special assumptions are made in the space X, and none on A except for (1.3) and (1.4). In particular the generation theorem applies to certain degenerate parabolic equations for which good *a priori* estimates (useful for proving existence assertions by compactness arguments) are not available: see example (a) in Section 2.

Let us also remark here that the generation theorem has several extensions. First, instead of (IVP) we may consider the evolution equation

$$\text{(IVP)}' \quad \begin{cases} \dfrac{du(t)}{dt} + A(t)(u(t)) = f(t) & 0 < t < \infty \\ u(0) = x_0 \end{cases}$$

governed by a t-dependent family $A(t)$ of m-accretive operators; under a hypothesis that the $A(t)$ have an "L^1 modulus of continuity in t" a convergence assertion similar to the generation theorem is proved in [28]. (See also Crandall-Pazy [27]). (In fact under a stronger assumption, (IVP)' can be directly treated as a special case of (IVP). This is accomplished by the ODE device of rewriting (IVP)' as

$$\frac{d}{dt}\binom{u(t)}{t} + A\binom{u(t)}{t} = \binom{f(t)}{0}$$

for the operator $A\binom{u}{t} \equiv \binom{A(t)u}{-1}$ on $Y \equiv X \times \mathbb{R}^1$; see [31] for details.)

Another version of the generation theorem covers the case that A is a multivalued m-accretive operator, in which case (IVP) reads

$$\text{(IVP)}'' \quad \begin{cases} \frac{du(t)}{dt} + A(u(t)) \ni f(t) & 0 < t < \infty \\ u(0) = x_0. \end{cases}$$

Extensions of the definitions and proofs to this case are straightforward. This degree of generality will be important for applications to example (b) in Section 2.

In view of the generation theorem the function $u(t)$ is an obvious candidate for our solution to (IVP). By construction $u(0) = x_0$, and so (1.2) is verified. As for additional properties of u and, in particular, the solvability of (1.2), we have these assertions:

Regularity Theorem (i) $u(t) \in C([0,T]; X)$; and $u(t)$ is Lipschitz continuous into X, if $f(t)$ is Lipschitz and $x_0 \in D(A)$.

(ii) (Brezis-Pazy [16]). If (IVP) has a strong solution $v(t)$ (i.e. if $v(t)$ is differentiable into X, $v(t) \in D(A)$, and v solves (1.1), (1.2) for a.e. t), then $u \equiv v$.

(iii) ([25], [44], [28]). If u is differentiable at some $t_0 \geq 0$ and if t_0 is a Lebesgue point of f, then $u(t_0) \in D(A)$ and $\frac{du(t_0)}{dt} + A(u(t_0)) = f(t_0)$.

(iv) If $x_0 \in D(A)$, and if either X is reflexive and f is of bounded variation or A is linear and $f \in C^1([0,T];X)$, then u is a strong solution of (IVP) on $[0,T]$.

Lest the reader be misled we now list some pathologies which may occur (for a nonlinear A in a nonreflexive space) even for $f \equiv 0$:

(v) (Crandall-Liggett [25]). $u(t) = S(t)x_0$ need not be differentiable for any $t > 0$ or any choice of $x_0 \in \overline{D(A)}$.

(vi) (Webb [55], Plant [48], Crandall [20]) even if $x_0 \in D(A)$, $u(t) = S(t)x_0$ need not belong to $D(A)$ for any $t > 0$.

Because of these possibilities $u(t)$ can in general be interpreted as a solution of (IVP) only in some weak sense;

see Benilan [4] for an explanation of how to accomplish this using the notion of so-called "good" solutions.

A recent result of Lê is interesting in light of (v) and (vi):

Theorem (Lê[41]). Let $\Omega \subseteq \mathbb{R}^n$. Suppose that A is an m-accretive operator on $L^1(\Omega)$, and that A restricted to the domain $\{x \mid x \in D(A) \cap L^\infty(\Omega),\ Ax \in L^\infty(\Omega)\}$ is also accretive in $L^\infty(\Omega)$. Let $u(t) = S(t)x_0$ denote the semigroup generated by A in $L^1(\Omega)$. Then

(1) If $x_0 \in D(A)$, $u(t) \in D(A)$ for a.e. $t > 0$.

(2) u is differentiable a.e. into $L^1(\Omega)$ and

$$\frac{du(t)}{dt} + Au(t) = 0 \text{ a.e.}$$

Thus u is a strong solution of (IVP) (for $f \equiv 0$). Lê bases her proof on the observation that the above hypotheses imply that A is accretive in $L^p(\Omega)$ for all $1 \leq p \leq \infty$, and in fact also in various intermediate Orlicz spaces.

The next theorem assures us that the "solution" u of (IVP) constructed by the generation theorem is well behaved with respect to small changes of x_0, f, and A:

Perturbation Theorem (cf. Benilan [4]). Let A and A^n $(n = 1,2,\ldots)$ be m-accretive operators; f and $f^n \in L^1(0,T;\ X)$ for some $T > 0$ $(n = 1,2,\ldots)$; and $x_0 \in \overline{D(A)}$; $x_0^n \in \overline{D(A^n)}$ $(n = 1,2,\ldots)$. Denote by $u(t)$ (resp. $u^n(t)$) the function constructed in the generation theorem corresponding to A, f, and x_0 (resp. A^n, f^n, x_0^n). If

(i) $x_0^n \to x_0$ in X,

(ii) $f^n \to f$ in $L^1(0,T;\ X)$, and

(iii) $A^n \to A$ (in the sense that for all $\lambda > 0$, $x \in X$

$$(I + \lambda A^n)^{-1}x \to (I + \lambda A)^{-1}x)$$

then
$$u^n \to u \text{ in } C([0,T],X).$$

Hence (IVP) is well-posed.

The construction of the semigroup can also be accomplished in various ways other than by the direct application of the generation theorem. A good general theorem of this type is the

Nonlinear Chernoff Formula (Brezis-Pazy [17]). Let A be an m-accretive operator and $S(t)$ the semigroup generated by A. Suppose that $F(t)$ $(t > 0)$ is a family of operators defined everywhere on X satisfying

(i) $\|F(t)x - F(t)\hat{x}\| \leq \|x - \hat{x}\|$ for all $t > 0$, $x,\hat{x} \in X$

and

(ii) $\frac{x - F(t)x}{t} \to A(x)$ as $t \searrow 0$, for each $x \in D(A)$.

Then

(1.6) $\lim_{n\to\infty} F(t/n)^n x_0 = S(t)x_0$ for each $t > 0$, $x_0 \in \overline{D(A)}$.

We shall describe applications of this formula in examples (a) and (c), Section 2.

Let us finally introduce some notation that provides an alternative characterization of accretiveness (often easier to verify in practice than (1.3)). For $x,y \in X$, define

$$[x,y]_+ \equiv \inf_{\lambda>0} \frac{\|x + \lambda y\| - \|x\|}{\lambda}$$

and

$$[x,y]_- \equiv \sup_{\lambda<0} \frac{\|x + \lambda y\| - \|x\|}{\lambda}.$$

It is easy to check that A is accretive if and only if

(1.3)' $0 \leq [x - \hat{x}, A(x) - A(\hat{x})]_+$ for all $x,\hat{x} \in D(A)$.

If A satisfies the stronger assumption

$$0 \leq [x - \hat{x}, A(x) - A(\hat{x})]_- \text{ for all } x,\hat{x} \in D(A),$$

A is called strongly accretive (a.k.a. accretive in the sense of Browder). If A is m-accretive and also either continuous or linear, then A is strongly accretive (see Barbu [3]).

The advantage of the alternative characterization (1.3)' of accretiveness is that for certain spaces X the brackets $[\,,]_\pm$ are easy to compute:

Theorem (cf. Sato [51]). Let $\Omega \subset \mathbb{R}^n$.

(i) if $X = L^p(\Omega)$ for $1 < p < \infty$

$$(1.7)\quad [f,g]_+ = [f,g]_- = \begin{cases} \frac{1}{\|f\|^{p-1}} \int |f|^{p-1} \operatorname{sgn} f \cdot g dx & \underline{f \not\equiv 0} \\ \|g\| & \underline{f \equiv 0}. \end{cases}$$

(ii) <u>if $X = L^1(\Omega)$</u>,

$$(1.8)\quad [f,g]_\pm = \int_{\{f>0\}} g\,dx - \int_{\{f<0\}} g\,dx \pm \int_{f=0} |g|dx;$$

(iii) <u>if $X = C(\overline{\Omega})$</u>,

$$(1.9)\quad [f,g]_+ = \max_{|f(x_0)|=\|f\|} g(x_0)\cdot \operatorname{sgn} f(x_0) \quad \underline{f \not\equiv 0},$$

$$(1.10)\quad [f,g]_- = \min_{|f(x_0)|=\|f\|} g(x_0)\cdot \operatorname{sgn} f(x_0) \quad \underline{f \not\equiv 0},$$

$$(1.11)\quad [f,g]_+ = [f,g]_- = \|g\| \quad \underline{f \equiv 0}.$$

A similar representation holds for $X = L^\infty(\Omega)$.

We close this section by noting that the recent book of Barbu [3] contains proofs of many of the results mentioned above.

2. Applications to Partial Differential Equations

In contrast to the basic theory presented in Section 1, the results described below are more recent and represent in some instances work still in progress. For this reason and also for heuristic purposes we state few exact theorems, and instead will emphasize the <u>formal</u> calculations indicating the applicability of the general theory and additional regularity properties. Furthermore we will for simplicity consider only cases in which there is no inhomogeneous term (i.e. $f \equiv 0$ in (IVP)).

a. The Porous Medium and Related Equations

<u>1</u>. We consider first the nonlinear partial differential equation

$$(2.1)\quad \begin{cases} u_t(x,t) - \Delta\varphi(u(x,t)) = 0 & (x,t)\in\Omega \times (0,\infty) \\ \varphi(u(x,t)) = 0 & (x,t)\in\partial\Omega \times (0,\infty) \\ u(x,0) = u_0(x) & x\in\Omega, \end{cases}$$

where $\varphi\colon \mathbb{R} \to \mathbb{R}$ is a nondecreasing function and Ω is a smooth domain in $\mathbb{R}^n$. If $\varphi(x) \equiv |x|^{\gamma-1}x$ for some $\gamma > 0$, (2.1) is called the <u>porous medium</u> equation. We note also that the classical Stefan problem can be converted (by the "entropy transformation") in the form (2.1); see Brezis [13].

Let us first prove that the nonlinear operator $A(u) \equiv -\Delta\varphi(u)$ (defined for u in some appropriate domain $D(A)$) is accretive, here in the space $X = L^1(\Omega)$. To simplify the discussion we will assume that φ is C^2 and is strictly increasing. Choose any two smooth functions u and $\hat{u}$, both vanishing on $\partial\Omega$. Let ρ be a bounded increasing function on $\mathbb{R}^1$, $0 = \rho(0)$. Then

$$(2.2) \qquad \begin{aligned} 0 &\leq \int |\nabla(\varphi(u) - \varphi(\hat{u}))|^2 \rho'(\varphi(u) - \varphi(\hat{u}))dx \\ &= \int (A(u) - A(\hat{u}))(\rho(\varphi(u) - \varphi(\hat{u}))dx. \end{aligned}$$

Now replace ρ by a sequence ρ_n with the same properties, $\rho_n(x) \to \mathrm{sgn}(x)$ for each $x \in \mathbb{R}$. Passing to limits in (2.2) we obtain, since $\mathrm{sgn}(\varphi(u) - \varphi(\hat{u})) = \mathrm{sgn}(u - \hat{u})$,

$$0 \leq \int A(u) - A(\hat{u}) \circ \mathrm{sgn}(u-\hat{u})dx \leq [u-\hat{u}, A(u)-A(\hat{u})]_+$$

by (1.8). According to (1.3)' the operator A, at least when restricted to a domain consisting of smooth functions vanishing on $\partial\Omega$, is accretive. With the larger domain $D(A) \equiv \{u \in L^1(\Omega) \mid \varphi(u) \in W_0^{1,1}(\Omega),\ \Delta\varphi(u) \in L^1(\Omega)\}$, A is still accretive and also satisfies the range conditions (1.4) (see Brezis-Strauss [18]). Hence, and even under no smoothness or strict monotonicity assumptions on φ, A generates a semi-group $S(t)u_0$ defined for each $t \geq 0$ and each $u_0 \in \overline{D(A)} = L^1(\Omega)$. All of the abstract theory concerning regularity, perturbations, etc., can now be invoked for the function $u(t) \equiv S(t)u$; but unfortunately, since L^1 is not reflexive, we cannot conclude that u has a derivative or that it solves (2.1) in any strong sense. Changing to another space does not help, since A is not accretive in $L^p(\Omega)$ $(1 < p \leq \infty)$ unless $\varphi(x) \equiv x$ (it is accretive in $H^{-1}(\Omega)$: see Brezis [13]). Some new assumptions and new methods are needed.

2. Explicit solutions of the porous medium equation show that in general $\frac{du}{dt}$ need not exist. But if φ^{-1} is Lipschitz (for a smooth function φ this is equivalent to assuming that

$\varphi' \geq \theta > 0$, i.e. that (2.1) is nondegenerate), $u(t)$ is a strong solution:

Theorem ([29]). Assume that φ is strictly increasing, $0 = \varphi(0)$, and φ^{-1} is Lipschitz continuous. Suppose also Ω is bounded. Then for each $u_0 \in L^1(\Omega)$, $u(t) \equiv S(t)u_0$ is differentiable a.e. into $L^2(\Omega)$ and so also into $L^1(\Omega)$, $u(t) \in D(A)$ a.e., and

$$\frac{du(t)}{dt} + A(u(t)) = 0 \text{ a.e. } [t], \tag{2.3}$$

with the estimate

$$\int_0^T \int_\Omega t^{(n+6)/2} u_t^2 \leq TC\|u_0\|^2_{L^1(\Omega)} \tag{2.4}$$

for each $t > 0$.

Hence u is a strong solution of (2.3), and even for a "bad" initial function $u_0 \notin D(A)$. Therefore $S(t)$ has a "smoothing effect" on the data, a phenomenon well known for semigroups in a Hilbert space governed by a subdifferential; the proof in this L^1 setting is in fact a modification of the Hilbert space argument. Note also that by statement (iii) in the regularity theorem, the second and third assertions of the theorem are automatic from the general theory once the differentiability into $L^1(\Omega)$ is proved.

For the porous medium equation, estimate (2.4) is false in general; but the semigroup in this case still displays various regularizing properties of a different kind:

Theorem. Suppose $\varphi(x) \equiv |x|^{\gamma-1}x$ for some $\gamma > 0$; and let $u(t) \equiv S(t)u_0$, $u_0 \in L^1(\Omega)$.

(i) (Veron [54], Benilan [5]). If $\gamma \geq 1$, then for each $t > 0$, $S(t)u_0 \in L^\infty(\Omega)$, with the estimate

$$\|u(t)\|_{L^\infty(\Omega)} \leq \frac{C}{t^{n/(2+(\gamma-1)n)}} \|u_0\|^{2/(2+(\gamma-1)n)}_{L^1(\Omega)} \tag{2.5}$$

(ii) ([23]). If $\gamma > 1$ and $m(\Omega) < \infty$, then

$$\|u(t)\|_{L^\infty(\Omega)} \leq \frac{C}{t^{1/(\gamma-1)}}, \tag{2.6}$$

where the constant C is independent of u_0.

(iii) ([23]). <u>If $0 < \gamma < 1$ and $m(\Omega) < \infty$, then for each $u_0 \in L^\infty(\Omega)$ there exists $T > 0$ such that</u>

$$u(t) = 0 \text{ for } t \geq T. \tag{2.7}$$

According to (i) the semigroup "smooths" integrable functions into bounded function ($\gamma \geq 1$); and by (ii) this effect is uniform with respect to u_0 if $m(\Omega) < \infty$. Estimates (2.5) and (2.6) can be derived <u>a priori</u> by a modification of Moser's iteration method for recursively estimating the L^p norms of a solution (cf. Veron [54]). Benilan [5] has proved results like (i) for a wider class of functions φ; his method is a clever adaptation of the DeGiorgi-Stampacchia techniques for elliptic equations in divergence form.

We should emphasize here that the preceding theorems are based upon certain formal, <u>a priori</u> estimates (like (2.4)-(2.6)) in the derivation of which the semigroup theory plays no role. The abstract considerations are useful, however, in constructing rigorous proofs once the formal estimates are in hand. For details of this see the references previously cited.

Benilan has recently proved that the semigroup solution is a strong solution of (2.1) for the case of porous medium equation in one space dimension; his arguments rely on Aronson's estimate ([1]) of $\max|(u^{\gamma-1})x|$. Various additional results for the porous medium equation in this case are to be found in Knerr [35].

<u>3</u>. A more direct application of the abstract theory occurs in the justification given of a numerical iteration scheme for solving (2.1); Berger, Brezis and Rogers [9].

Assume now that φ is Lipschitz continuous on bounded intervals, $\varphi(0) = 0$, and that $u_0 \in L^\infty(\Omega)$. We propose to approximate a solution of (2.1) by the iteration scheme

$$\begin{cases} \dfrac{u_k^n - u_{k-1}^n}{\lambda_n} + \dfrac{(I - T(\mu\lambda_n))}{\mu\lambda_n}\varphi(u_k^n) = 0 \quad k = 1,2,\ldots & (2.8) \\ u_0^n = u_0, & (2.9) \end{cases}$$

when μ is the Lipschitz constant for φ on $[-\|u_0\|_{L^\infty}, \|u_0\|_{L^\infty(\Omega)}]$, $\lambda_n > 0$, and $T(t)$ denotes the linear semigroup generated in $L^1(\Omega)$ by $\hat{A} = -\Delta$ (i.e. $u(x,t) \equiv T(t)u_0(x)$ is a solution of the heat equation in $\Omega \times (0,\infty)$). Define the step function $u^n(t) = u_k^n$ on $(k-1)\lambda_n \leq t \leq k\lambda_n$ (and contrast this to the definition of u^n in the generation theorem).

Theorem. ([9]). For each $t > 0$

$$u^n(t) \to u(t) = S(t)u_0 \text{ in } L^1(\Omega) \tag{2.10}$$

as $\lambda_n \searrow 0$.

Sketch of Proof. An induction argument shows that each u_k^n remains in the interval $[-\|u_0\|_{L^\infty}, \|u_0\|_{L^\infty}]$; and so, redefining φ off this interval if necessary, we may assume that φ has μ as a global Lipschitz constant. For each $t \geq 0$ and $u \in L^1(\Omega)$, define

$$F(t)u \equiv u - \frac{1}{\mu}[T(\mu t)\varphi(u) - \varphi(u)].$$

Then, by (2.8),

$$u_n(t) \equiv F(t/n)^n u_0.$$

Hypothesis (i) of the nonlinear Chernoff formula is not hard to check, and hypothesis (ii) is also verified since $\frac{u-F(t)u}{t} = \frac{\varphi(u)-T(\mu t)\varphi(u)}{\mu t} \to -\Delta\varphi(u) = A(u)$ for $u \in D(A)$. (A linear semigroup is differentiable at 0 for u in the domain of its generator). The theorem follows. ||

Notice, now that the converge of the approximation scheme (2.8), (2.9) is proved, the (semigroup) solution of the nonlinear, possibly degenerate p.d.e. (2.1) can be computed in terms of solutions to the linear heat equation.

b. Variational and Quasi-variational Equations of Evolution

1. Assume now that β is a possibly multivalued maximal monotone graph on $\mathbb{R}^1$ (i.e. β is a nondecreasing function

with the "jump filled in" at any point of discontinuity, c.f. Brezis [11]). We begin this section by studying the p.d.e.s

$$(2.11)\quad \begin{cases} u_t(x,t)-\Delta u(x,t) + \beta(u(x,t))\ni 0 & (x,t)\in\Omega \times (0,\infty) \\ u(x,t) = 0 & (x,t)\in\partial\Omega \times (0,\infty) \\ u(x,0) = u_0(x) & x\in\Omega \end{cases}$$

and

$$(2.12)\quad \begin{cases} u_t(x,t) - \Delta u(x,t) = 0 & (x,t)\in\Omega \times (0,\infty) \\ -\frac{\partial u}{\partial n}(x,t)\in\beta(u(x,t)) & (x,t)\in\partial\Omega \times (0,\infty) \\ u(x,0) = u_0(x) & x\in\Omega. \end{cases}$$

(" $\frac{\partial}{\partial n}$ " denotes the outward normal derivative). In the case that

$$(2.13)\qquad \beta(x) = \begin{cases} 0 & x > 0 \\ (-\infty,0] & x = 0, \\ \emptyset & x < 0 \end{cases}$$

these equations are evolutions governed by $-\Delta$, but subject to the constraint $u \geq 0$ in Ω and on $\partial\Omega$, respectively.

It is simple to verify that the formal operations $A_1(u) \equiv -\Delta u + \beta(u)$ (for functions u vanishing on $\partial\Omega$) and $A_2(u) \equiv -\Delta u$ (for functions u satisfying $-\frac{\partial u}{\partial n}\in\beta(u)$ on $\partial\Omega$) are accretive in $L^p(\Omega)$ for <u>all</u> $1 \leq p \leq \infty$. For $1 < p < \infty$ this results from (1.3)', the characterization (1.7) of the bracket $[\,,]_+$ in L^p, and an integration by parts; the cases $p = 1,\infty$ follow by passing to limits as $p \to 1,\infty$ in inequality (1.3). In fact with proper domains, depending on p, A_1 and A_2 are m-accretive (although possibly multivalued) in each $L^p(\Omega)$, and so generate semigroups $S_1^p(t)$ and $S_2^p(t)$ $(t \geq 0, 1 \leq p \leq \infty)$. The semigroups generated in L^1 in each case extend the others (assuming $m(\Omega) < \infty$), and these we denote simply by $S_1(t)$ and $S_2(t)$.

<u>2</u>. Since A_1 and A_2 can be realized in <u>every</u> space $L^p(\Omega)$ as m-accretive, we should expect the semigroup solutions

$$(2.14)\qquad u_1(t) \equiv S_1(t)u_0,\ u_2(t) \equiv S_2(t)u_0$$

of (2.11) and (2.12) to have good regularity properties. Indeed if $u_0\in D_p(A_i)$ (= domain of A_i in $L^p(\Omega)$) for some

$1 < p < \infty$, then statement (iv) of the regularity theorem implies at once that $u_i(t)$ is a strong solution. And in fact this is the case even for merely integrable initial data u_0:

Theorem (Massey [43], [30]). Assume $u_0 \in L^1(\Omega)$, and that the functions $u_i(t)$ are defined by (2.14) (i = 1,2.). Suppose that $0 \in \beta(0)$. Then $u_i(t)$ is differentiable a.e. and is a strong solution of $\frac{du_i}{dt} + A_i(u_i(t)) \ni 0$ (i = 1,2) in $L^p(\Omega)$ $(1 \leq p < \infty)$, with the estimates

$$(2.15) \qquad \|u_i(t)\|_{L^\infty(\Omega)} \leq \frac{C}{t^{n/2}} \|u_0\|_{L^1(\Omega)} \quad (i = 1,2)$$

and

$$(2.16) \qquad \|\frac{d}{dt}u_i(t)\|_{L^\infty(\Omega)} \leq \frac{C}{t^{(n+2)/2}}\|u_0\|_{L^1(\Omega)} \quad (i = 1,2)$$

for $0 < t \leq T$. (In the case i = 1, C does not depend on T).

Inequalities (2.15) and (2.16) can be derived formally from the iteration technique of Moser and the general estimate of Brezis ([11])

$$(2.17) \qquad \|\frac{du_i}{dt}(t)\|_{L^2(\Omega)} \leq \frac{C}{t}\|u_i(t/2)\|_{L^2(\Omega)} \qquad (i = 1,2).$$

Another proof is based on a device of Massey [43], which we describe here for the case i = 1. Let us compare a solution u of (2.11) to the auxillary functions $v^{\pm}$ solving the linear problems

$$\begin{cases} v^{\pm}_t(x,t) - \Delta v^{\pm}(x,t) = 0 & (x,t) \in \Omega \times (0,\infty) \\ v^{\pm}(x,t) = 0 & (x,t) \in \partial\Omega \times (0,\infty) \\ v^{+}(x,0) = \max(u_0(x),0), v^{-}(x,0) = \min(u_0(x),0) & x \in \Omega. \end{cases}$$

An easy argument proves

$$v^{-}(x,t) \leq u(x,t) \leq v^{+}(x,t) \qquad (x,t) \in \Omega \times (0,\infty);$$

and therefore for $t > 0$

$$\begin{aligned} \|u(\cdot,t)\|_{L^\infty} &\leq \|v^{-}(\cdot,t)\|_{L^\infty} + \|v^{+}(\cdot,t)\|_{L^\infty} \\ &\leq \frac{C}{t^{n/2}}(\|v^{-}(\cdot,0)\|_{L^1} + \|v^{+}(\cdot,0)\|_{L^1}) \\ &= \frac{C}{t^{n/2}}\|u_0\|_{L^1}. \end{aligned}$$

An analagous argument applied to difference quotients gives various estimates on $\|u_t\|_{L^\infty}$, and these along with (2.17) lead to (2.16).

For the case i = 2, we compare a solution of (2.12) to functions $w^{\pm}$ solving the heat equation with Neumann boundary conditions.

Additional estimates are possible for (2.11) in certain special cases, for example (2.13). In this situation we approximate β by a sequence of smooth, concave functions β_n and then consider an auxillary function

$$v^n \equiv \zeta^2 (u^{n\,-}_{\xi\xi})^2 + \lambda |\nabla u^n|^2 - \mu t,$$

where u^n solves (2.11) with β_n replacing β, $\zeta = \zeta(x,t)$ is cutoff function, $u^n_{\xi\xi}$ denotes a pure second derivative in some direction ξ, and λ and μ are constants chosen so large that v^n cannot have an interior maximum in $\Omega \times (0,\infty)$. This is accomplished by estimates similar to those described in Ladyženskaja et al [40, p. 414-416]. We thus derive a bound from below on any pure second derivative; and this leads in turn to the estimate

$$(2.18) \qquad \|u_{x_i x_j}\|_{L^\infty(Q)} \leq C(Q) \qquad (i,j = 1,\ldots,n)$$

for any $Q \subset\subset \Omega \times (0,\infty)$. (When $-\Delta$ is replaced in (2.11) by an elliptic second order operation with coefficients depending on x, the method just described does not work, but inequality (2.18) can still be proved by a parabolic version of the techniques in Brezis-Kinderlehrer [15]).

It is worth remarking here that such good estimates are available for problems (2.11) and (2.12) that existence can be proved more or less directly by compactness arguments, without recourse to the Crandall-Liggett existence theorem.

3. As a next example of a variational inequality let us consider an evolution equation with a constraint on u_t (cf. Brezis [12], Lions [42], Konishi [36]):

$$(2.19)\quad \begin{cases} \gamma(u_t(x,t)) - \Delta u(x,t) \ni 0 & (x,t)\in\Omega \times (0,\infty) \\ u(x,t) = 0 & (x,t)\in\partial\Omega \times (0,\infty) \\ u(x,0) = u_0(x) & x\in\Omega, \end{cases}$$

where γ is a maximal monotone graph on $\mathbb{R}^1$. We set $\beta(x) \equiv -\gamma^{-1}(-x)$ to rewrite the first equation above as

$$(2.19)'\quad u_t(x,t) + \beta(-\Delta u(x,t)) \ni 0 \qquad (x,t)\in\Omega \times (0,\infty).$$

Now from (1.3)' and (1.9) we note that

(2.20) <u>if A is an accretive operation on $X = C(\bar{\Omega})$ (or $L^\infty(\Omega)$), then $B(u) \equiv \beta(A(u))$ is accretive, for any maximal monotone graph β on $\mathbb{R}^1$.</u> (<u>Here $D(B) = \{u\in D(A) \mid A(u)\in D(\beta) \text{ a.e.}\}$</u>).

This observation applies to the case at hand, and so, formally at least, (2.19) falls within the general semigroup framework. For details as to the precise applicability of the general theory, we refer the interested reader to Benilan-Ha [7].

We note in passing that <u>if</u> γ <u>and</u> γ^{-1} are single-valued and Lipschitz on $\mathbb{R}^1$, the Moser-Nash and the Schauder parabolic estimates can be invoked to derive the <u>a priori</u> estimate

$$\|u\|_{C^{2+\alpha,1+\alpha/2}(Q)} \leq C(Q)$$

for some $0 < \alpha < 1$ and each $Q \subset\subset \Omega \times (0,\infty)$.

<u>4</u>. Next consider the <u>quasi</u>-variational inequality of evolutions

$$(2.21)\quad \begin{cases} u_t(x,t) - \Delta u(x,t) \geq 0,\ u(x,t) \geq M(u)(x,t) & (x,t)\in\Omega \times (0,\infty) \\ (u_t - \Delta u)(u - M(u)) = 0 \text{ a.e.} & \\ u(x,t) = 0 & (x,t)\in\partial\Omega \times (0,\infty) \\ u(x,0) = u_0(x) & x\in\Omega, \end{cases}$$

where M is some operator on $L^\infty(\Omega)$ which satisfies

$$(2.22)\qquad \|M(u) - M(\hat{u})\|_{L^\infty} \leq \|u - \hat{u}\|_{L^\infty} \text{ for all } u,\hat{u}\in L^\infty(\Omega).$$

We may rewrite the first part of (2.21) as

$$(2.21)' \quad u_t(x,t)-\Delta u(x,t) + \beta(u(x,t)-M(u)(x,t)) \ni 0 \qquad (x,t)\in\Omega \times (0,\infty)$$

for β defined by (2.13). By (2.22), (1.5) and (2.20) the <u>formal</u> operator $\beta(u-M(u))$ is accretive in $L^{\infty}(\Omega)$ and so also is $A(u) \equiv -\Delta u + \beta(u-M(u))$ (with $u = 0$ on $\partial\Omega$ for functions in its domain).

Note carefully that these calculations are only formal and merely indicate the applicability of the abstract theory to (2.21). The task of determining a precise domain for which which A satisfies (1.4) has not been accomplished.

The preceding observations are due to Ph. Benilan.

<u>5</u>. A similar problem arising in the study of plasma physics is this:

$$(2.23) \quad \begin{cases} u_t(x,t)-\Delta u(x,t) + \text{mes}\{y\in\Omega \mid u(y,t) \le u(x,t)\} = 0 & (x,t)\in\Omega \times (0,\infty) \\ u(x,t) = 0 & (x,t)\in\partial\Omega \times (0,\infty) \\ u(x,0) = u_0(x) & x\in\Omega. \end{cases}$$

The attentive reader should by now be able to carry out the formal calculations indicating that $A(u) \equiv -\Delta u + \text{mes}\{y \mid u(y) \le u(x)\}$ is accretive in $C(\overline{\Omega})$ or $L^{\infty}(\Omega)$. For details of the semigroup approach to (2.23) see Benilan [6] (cf. also Mossino [45]).

c. <u>Bellman's Equation of Dynamic Programming</u>

<u>1</u>. According to heuristic reasoning the optimal payoff $u(x,t)$ of a certain stochastic differential system (starting at time t in state $x\in\overline{\Omega}$) satisfies the nonlinear parabolic p.d.e.

$$(2.24) \quad \begin{cases} u_t(x,t) + \sup_{\gamma\in\Gamma}[A^{\gamma}u(x,t)] = 0 & (x,t)\in\Omega \times (0,\infty) \\ u(x,t) = 0 & (x,t)\in\partial\Omega \times (0,\infty) \\ u(x,0) = u_0(x) & x\in\Omega \end{cases}$$

Here Γ is the set of admissible actions; and, for each $\gamma\in\Gamma$,

A^γ is a second order linear elliptic operator of the form

(2.25) $A^\gamma v \equiv -a^\gamma_{ij}(x)v_{x_i x_j} + b^\gamma_i(x)v_{x_i} + c^\gamma(x)v$

(where $\theta|\xi|^2 \leq a^\gamma_{ij}(x)\xi_i\xi_j \leq \Theta|\xi|^2$ for all $\xi\in\mathbb{R}^n$ and $\Theta \geq \theta > 0$), characterizing the response of the system to the constant control γ. See Fleming-Rishel [33], Kushner [38], or Bensoussan-Lions [8] for the formal derivation of (2.24).

The applicability of abstract semigroup theory to (2.24) was first noted by Pliska [50], who used probabilistic arguments to show that the operator $\sup_\gamma[A^\gamma u]$ is accretive in $X = C(\bar{\Omega})$ (or $L^\infty(\Omega)$). We indicate here another proof by observing

(2.26) $\left\{\begin{array}{l}\text{if } A^\gamma\ (\gamma\in\Lambda) \text{ is any collection of strongly accretive} \\ \text{operators on } X = C(\bar{\Omega}), \text{ then the operators} \\ A(u) \equiv \sup_{\gamma\in\Gamma}[A^\gamma(u)] \text{ and } B(u) \equiv \inf_{\gamma\in\Gamma}[A^\gamma(u)] \text{ are also} \\ \text{strongly accretive (for} \\ D(A) \equiv \{u\in\cap_\gamma D(A^\gamma)\ A(u)\in C(\bar{\Omega})\} \text{ and } D(B) \text{ defined simi-} \\ \text{larly)}\end{array}\right.$

To establish (2.26) we employ the characterization (1.10) of the bracket $[\,,]_-$ in $C(\bar{\Omega})$. Choose $u,\hat{u}\in \bigcap_{\gamma\in\Gamma} D(A^\gamma)$, and let x_0 be a point in Ω when (WLOG) $u(x_0)-\hat{u}(x_0) = \|u-\hat{u}\|_{C(\bar{\Omega})}$. According to (1.3)', (1.10), and the strong accretiveness of each A^γ, we have

$$A^\gamma(u)(x_0) - A^\gamma(\hat{u})(x_0) \geq 0 \text{ for all } \gamma\in\Gamma.$$

This implies $\sup_{\gamma\in\Gamma}[A^\gamma(u)](x_0) - \sup_{\gamma\in\Gamma}[A^\gamma(\hat{u})](x_0) \geq 0$ and so

$$0 \leq [u-\hat{u}, A(u)-A(\hat{u})]_-,$$

i.e., A is strongly accretive. The proof for B is similar. Note that the operators defined by (2.25) satisfy the hypothesis of (2.26) if $c^\gamma(x) \geq 0$.

2. Now that we have shown A defined by (2.26) is accretive in $C(\bar{\Omega})$, we must verify the range condition (1.4); that is, we must solve the stationary problem

$$(2.27) \qquad \begin{cases} u(x) + \sup_{\gamma\in\Gamma}[A^{\gamma}u(x)] = f(x) & x\in\Omega. \\ u(x) = 0 & x\in\partial\Omega. \end{cases}$$

for all $f\in C(\overline{\Omega})$. As this equation is not even quasilinear (unless $a_{ij}^{\gamma}(x)$ is independent of γ for all $\gamma\in\Gamma$), there are no standard theorems from the p.d.e. literature which apply. The solvability of (2.27) for an arbitrary smooth domain Ω remains an open question, but certain partial results are known.

Under various reasonable hypotheses on the smoothness of the coefficient of A^{γ} with respect to x and γ, N. V. Krylov [37] has constructed a solution $u\in W_{loc}^{2,p}(\mathbb{R}^n)$ $(1 \leq p < \infty)$ for (2.27) in the case that $\Omega = \mathbb{R}^n$, $f\in C^2(\mathbb{R}^n)$. For a bounded domain Ω considerably less is known:

Theorem. Let Ω be a bounded smooth domain in $\mathbb{R}^n$. Assume that for each $\gamma\in\Gamma$, a_{ij}^{γ}, b_i^{γ}, $c^{\gamma}\in C^2(\overline{\Omega})$. Suppose $c^{\gamma}(x) \geq \mu$ for some appropriate constant μ. Then

(i) if $n = 2$ and $f\in L^2(\Omega)$, there exists a unique $u\in H^2(\Omega) \cap H_0^1(\Omega)$ solving (2.29). If $f\in W^{1,p}(\Omega)$ for some $p > 2$, $u\in C^{2,\alpha}(\Omega)$ for some $\alpha > 0$.

(ii) ([14]) if n is arbitrary, but Γ consists of only two elements, (that is, if (2.27) reads

$$(2.27)' \qquad u + \max_{i=1,2} [A^i u] = f)$$

there exists a unique solution $u\in H^2(\Omega) \cap H_0^1(\Omega)$ for each $f\in L^2(\Omega)$. If $f\in W^{1,p}(\Omega)$ for $p > n$, then $u\in H^3(\Omega) \cap C(\overline{\Omega}) \cap C^{2,\alpha}(\Omega)$ for some $0 < \alpha < 1$. Statement (i) is relatively simple to prove by the continuation of parameter method, using the good *a priori* elliptic estimates available in 2 dimensions: see Ladyženskaja-Ural'ceva [39, §3.17]. The proof in [14] uses variational inequality methods to establish existence, and the DeGiorgi-Moser-Stampacchia and the Schauder estimates to prove interior regularity.

3. In those cases for which (2.27) is solvable the generation theorem applies to give a semigroup solution to (2.24).

However - here we continue some remarks in part(b)-for the case of only two operators such good a priori estimates are available for (2.24) that there is no need to resort to the semigroup theory. Indeed, it is not hard to modify the existence theorem proof of [14] for a solution of (2.27)' to solve directly the parabolic problem (2.24) (where $\Gamma = \{1,2\}$). Furthermore, the function $u(x,t)$ so obtained belongs to the space $C^{2+\alpha,1+\alpha/2}(Q)$ $(Q = \Omega \times (0,T))$ for some $0 < \alpha < 1$, and thus is a classical solution.

4. In the more interesting cases that Γ comprises more than two elements these regularity proofs do not work. In these situations the notion of semigroup is still useful.

M. Nisio has constructed [46] a semigroup solution of (2.24) in the case $\Omega = \mathbb{R}^n$ by a method quite different from that discussed in Section 1; see also Bensoussan-Lions [8] for an exposition. In a second paper Nisio [47] constructs in an abstract setting - again without using any motion of accretiveness or using the generation theorem - a semigroup solution of

$$(2.28)\qquad \begin{cases} \dfrac{du}{dt} + \sup\limits_{\gamma\in\Gamma}[A^{\gamma}(u)] = 0 & 0 < t < \infty \\ u(0) = u_0 \end{cases}$$

where the A^{γ} are linear semigroup generators in $X = L^{\infty}(\Omega)$, satisfying various additional hypotheses (which do not seem to preclude application to the case (2.25)). Her method is this:

First, let $S^{\gamma}(t)$ denote the linear semigroup on $L^{\infty}(\Omega)$ generated by A^{γ}. Next set

$$(2.29)\qquad F(t)u_0 \equiv \inf_{\gamma\in\Gamma}[S^{\gamma}(t)u_0].$$

Nisio then proves that $\lim\limits_{n\to\infty} F(t/n)^n u_0 \equiv S(t)u_0$ exists and that $S(t)$ is the required semigroup.

We emphasize that all of this is carried out with no reference to abstract theory discussed in Section 1. However (following an observation of Brezis) let us note here that the convergence of $F(t/n)^n u_0$ would be a consequence of the nonlinear Chernoff formula <u>if it were known</u> that $A(u) \equiv \max_{\gamma\in\Gamma}[A^\gamma(u)]$ is m-accretive (the accretiveness is automatic by (2.26)). Indeed

$$\lim_{t\searrow 0} \frac{u-F(t)u}{t} = \lim_{t\searrow 0} \sup_{\gamma\in\Gamma} \frac{u_0-S^\gamma(t)u_0}{t}$$
$$= \sup_{\gamma\in\Gamma}[A^\gamma u] = A(u)$$

for $u\in \bigcap_{\gamma\in\Gamma} D(A^\gamma) \equiv D(A)$. Thus $F(t/n)^n u_0 \to S(t)u_0$ where $S(t)$ is the semigroup generated by the (assumed) m-accretive operator A.

We stress that this observation does not trivialize Nisio's work (since she proved convergence without supposing that A satisfies a range condition); but it does make clearer the connections between her approach and the abstract theory presented here.

There are clearly many open questions concerning the stationary and evolution Bellman equation, and much work yet to be done.

<u>5</u>. Isaac's equation from differential game theory (see Friedman [34]) is a generalization of (2.24):

$$\begin{cases} u_t(x,t) + \sup_{\gamma\in\Gamma} \inf_{\delta\in\Delta}[A^{\gamma,\delta}u(x,t)] = 0 & (x,t)\in\Omega \times (0,\infty) \\ u(x,t) = 0 & (x,t)\in\partial\Omega \times (0,\infty) \\ u(x,0) = u_0(x) & x\in\Omega, \end{cases}$$

where the $A^{\gamma,\delta}$ are second order elliptic operators. According to (2.26)

$$A(u) \equiv \sup_{\gamma\in\Gamma} \inf_{\delta\in\Delta}[A^{\gamma,\delta}u]$$

is accretive in $C(\overline{\Omega})$. Hence the semigroup theory should apply, but beyond this simple observation nothing is known.

d. <u>Other Applications</u>

We conclude with some bibliographical references to

other applications of the abstract semigroup theory to non-linear differential equations:

1. Single nonlinear conservation law - Crandall [21]
2. Hamilton-Jacobi equations - Aziwa [2], Tamburro [53], Burch [19].
3. Delay differential and integro-differential equations, Webb [56], Plant [49], Crandall-Nohel [26], Flaschka-Leitman [32], Bressen-Dyson [10].
4. Control theory, Slemrod [52].

References

1. D. G. Aronson, Regularity properties of flows through porous media, SIAM J. Appl. Math. 17 (1969), 461-467.

2. S. Aizawa, A semigroup treatment of the Hamilton-Jacobi equation in one space variable, Hiroshima Math. J. 3 (1973), 367-386.

3. V. Barbu, Nonlinear Semigroups and Differential Equations in Banach Spaces, Noordhoff International Publishing, 1976.

4. Ph. Benilan, Équations d'évolution dans un espace de Banach quelcongue et applications, thesis, Orsay, 1972.

5. __________, Operateurs accretifs et semigroupes dans les espaces L^p $(1 \leq p \leq \infty)$, to appear.

6. __________, to appear.

7. __________ and K. Ha, Équation d'évolution du type $(du/dt) + \beta(\partial\varphi(u)) \ni 0$ dans $L^\infty(\Omega)$, C. R. Acad. Sc. 281 (1975), 947-950.

8. A. Bensoussan and J. L. Lions, Contrôle optimal par temps d'arrêt et contrôle impulsionnel, Hermann, Vol. I, 1977.

9. A. Berger, H. Brezis, J. Rogers, to appear.

10. R. V. Bressen and J. Dyson, Functional differential equations and nonlinear evolution operators, Proc. Royal Soc. (Edinburgh) 5A, 20 (1975) 223-234.

11. H. Brezis, Opérateurs maximaux monotones et semigroupes de contractions dans les espaces de Hilbert, North Holland, 1973.

12. ________, Problèmes unilatéraux, J. Math. Pures Appl. 51 (1972), 1-164.

13. ________, On some degenerate nonlinear parabolic equations, Proc. Symp. Pure Math., Vol. 18, AMS, 1970.

14. ________ and L. C. Evans, A variational inequality approach to the Bellman-Dirichlet equation for two elliptic operators, to appear in Arch. Rat. Mech. Anal.

15. ________ and D. Kinderlehrer, The smoothness of solutions of nonlinear variational inequalities, Indiana U. Math. J. 23 (1974), 831-844.

16. ________ and A. Pazy, Accretive sets and differential equations in Banach spaces, Israel J. Math. 8 (1970), 367-383.

17. ________ and A. Pazy, Convergence and approximation of semigroups of nonlinear operators in Banach spaces, J. Func. Anal. 9 (1972), 63-74.

18. ________ and W. Strauss, Semilinear elliptic equations in L^1, J. Math. Soc. Japan 25 (1973), 565-590.

19. C. Burch, A semigroup treatment of the Hamilton-Jacobi equation in several space variables, J. Diff. Eq. 23 (1977), 107-124.

20. M. G. Crandall, A generalized domain for semigroup generators, Proc. AMS. 37 (1973), 434-439.

21. ____________, The semigroup approach to first order quasilinear equations in several space variables, Israel J. Math. 12 (1972), 108-132.

22. ____________ and L. C. Evans, On the relation of the operator $\frac{\partial}{\partial s} + \frac{\partial}{\partial \tau}$ to evolution governed by accretive operators, Israel J. Math. 21 (1975), 261-278.

23. ____________ and L. C. Evans, to appear.

24. ____________ and T. Liggett, Generation of semi-groups of nonlinear transformations on general Banach spaces, Amer. J. Math. 93 (1971), 265-298.

25. ______________ and T. Liggett, A theorem and a counter-example in the theory of semigroups of nonlinear trans formations, Trans. AMS 160 (1971), 263-278.

26. ______________ and J. Nohel, An abstract functional differential equation and a related nonlinear Volterra equation, to appear.

27. ______________ and A. Pazy, Nonlinear evolution equations in Banach spaces, Israel J. Math. 11 (1972), 57-94.

28. L. C. Evans, Nonlinear evolution equations in an arbitrary Banach space, Israel J. Math. 26 (1977), 1-41.

29. ___________, Differentiability of a nonlinear semigroup in L^1, J. Math. Anal. and Appl. 60 (1977), 703-715.

30. ___________, Regularity properties for the heat equation subject to nonlinear boundary constraints, to appear in J. Nonlinear Anal.

31. ___________ and F. J. Massey III, A remark on the construction of nonlinear evolution operators, Houston J. Math 3 (1977).

32. H. Flaschka and M. J. Leitman, On semigroups of nonlinear operators and the solution of the functional differential equation $x'(t) = F(x_t)$, J. Math. Anal and Appl. 49 (1975), 649-658.

33. W. H. Fleming and R. W. Rishel, <u>Deterministic and Stochastic Optimal Control</u>, Springer, 1975.

34. A. Friedman, <u>Differential Games</u>, Wiley-Interscience, 1971.

35. B. Knerr, Some results concerning solutions of the Cauchy problem for the porous medium equation when the initial data have compact support, to appear in Trans. AMS.

36. Y. Konishi, On the nonlinear semigroups associated with $u_t = \Delta\beta(u)$ and $\varphi(u_t) = \Delta u$, J. Math. Soc. Japan 25 (1973), 622-628.

37. N. V. Krylov, Control of a solution of a stochastic integral equation, Theory Prob. and Appl. 17 (1972), 114-131.

38. H. J. Kushner, Stochastic Stability and Control, Academic Press, 1967.

39. O. A. Ladyženskaja and N. N. Ural'ceva, Linear and Quasilinear Elliptic Equations, Academic Press, 1968.

40. O. A. Ladyženskaja, V. A. Solonnikov, and N. N. Ural'ceva, Linear and Quasilinear Equations of Parabolic Type, AMS, 1968.

41. Lê C. H., Dérivabilité d'un semi-groupe engendré par un opérateur m-accretif de $L^1(\Omega)$ and accretif dan $L^\infty(\Omega)$, C. R. Acad. Sc. 283 (1976), 469-472.

42. J. L. Lions, Sur quelques questions d'analyse, de mécanique, et de contrôle optimal, University of Montreal Press, 1976.

43. F. J. Massey III, Semilinear parabolic equations with L^1 initial data, Indiana U. Math. J. 26 (1977), 399-411.

44. I. Miyadera, Some remarks on semi-groups of nonlinear operators, Tôhoku Math. J. 23 (1971), 245-258.

45. J. Mossino, Sur certaines inéquations quasivariationnelles apparaissant en physique, C. R. Acad. Sc. 282 (1976), 187-190.

46. M. Nisio, Remarks on stochastic optimal controls, Japan J. Math 1 (1975), 159-183.

47. M. Nisio, On a non-linear semi-group attached to stochastic optimal control, to appear.

48. A. T. Plant, Flow-invariant domains of Hölder continuity for nonlinear semigroups, Proc. AMS 53 (1975), 83-87.

49. __________, Nonlinear semigroups of translations in Banach space generated by functional differential equations, to appear.

50. S. R. Pliska, A semigroup representation of the maximum expected reward vector in continuous parameter Markov decision theory, SIAM J. Control 13 (1975), 1115-1129.

51. K. Sato, On the generators of non-negative contraction semi-groups in Banach lattices, J. Math. Soc. Japan 20 (1968), 431-436.

52. M. Slemrod, An application of maximal dissipative sets in control theory, J. Math. Anal. and Appl. 46 (1974), 369-387.

53. M. B. Tamburro, The evolution operator solution of the Cauchy problem for the Hamilton-Jacobi equation, Israel J. Math. 26 (1977), 232-264.

54. L. Veron, Coercivite et proprietes regularisantes des semi-groupes non lineaires dans less espaces de Banach, to appear.

55. G. F. Webb, Nonlinear perturbations of linear accretive operators in Banach spaces, Israel J. Math. 12 (1972), 237-248.

56. __________, Autonomous nonlinear functional differential equations and nonlinear semigroups, J. Math. Anal. and Appl. 46 (1974), 1-12.

Supported in part by NSF Grant MCS77-01952.

Department of Mathematics
University of Kentucky
Lexington, Kentucky 40506

Finite Time Blow-Up in Nonlinear Problems

J. M. Ball

1. Introduction.

It is well known that solutions of ordinary differential equations may blow up in finite time; for example, $x(t) = \frac{1}{1-t}$ is a solution of the equation $\dot{x} = x^2$. Furthermore, for ordinary differential equations of the form

$$(1.1) \qquad \dot{x} = f(x,t), \quad x \in \mathbb{R}^n, \quad t \in \mathbb{R} ,$$

with $f:\mathbb{R}^n \times \mathbb{R} \to \mathbb{R}^n$ continuous, a standard existence and continuation theorem (cf Hartman [11]) asserts that finite time blow-up is equivalent to global nonexistence. More precisely, if $(x_0,t_0) \in \mathbb{R}^n \times \mathbb{R}$ there exists a solution $x(t)$ of (1.1) with $x(t_0) = x_0$, defined on a maximal interval of existence $[t_0,t_{max})$, where $t_0 < t_{max} \leq \infty$, and if $t_{max} < \infty$ then

$$\lim_{t \nearrow t_{max}} |x(t)| = \infty .$$

The situation for infinite-dimensional initial value problems, such as those arising from partial differential equations, is more complicated and it is not possible to make any general statement relating blow-up and nonexistence. In part this is due to the coexistence of nonequivalent norms each of which may serve as a measure for the size of a solution.

In recent years, much experience has been gained in the use of differential inequalities for the study of global nonexistence for infinite-dimensional problems. The reader is

ISBN 0-12-195250-9

referred to the papers [6], [8-9], [12-15], [18-21], [24], [29-30] for some of this work. The idea is to derive a differential inequality for a real valued functional $F(u(t))$ of the solution u of the problem under consideration. The inequality is then solved, subject to appropriate initial conditions at $t = t_0$, so as to obtain a lower bound for $F(u(t))$ that blows up at some finite time $t_1 > t_0$. If the definition of solution requires F to be finite for all time then global nonexistence has been established. It cannot in general be concluded, however, that $F(u(t))$ itself blows up at some finite time, since the maximal half-open interval of existence of the solution may be $[t_0, t_{max})$, where $t_{max} < t_1$. (See Figure 1.) An example of this phenomenon for a backwards nonlinear heat equation is described in Ball [1].

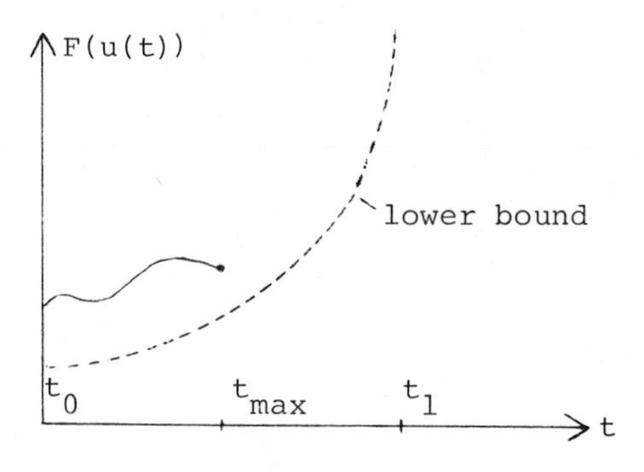

Figure 1.

Once it is known that $t_{max} < \infty$ then blow-up of u follows if a continuation theorem analogous to that described above for ordinary differential equations holds. Such a theorem will imply that <u>some</u> norm of u blows up, though not necesarily that $F(u)$ does. To obtain blow-up of $F(u)$ or other measures of u arguments based on more detailed structure of the problem may be needed.

The interpretation of blow-up theorems in physical problems often poses difficulties; blow-up may indicate either a real phenomenon or a failure of the physical model. In continuum mechanics, for example, hypotheses concerning the behaviour of constitutive functions for unbounded arguments may be necessary to establish blow-up, yet such values of the arguments may not be physically realistic. In such cases careful quantitative estimates are needed to decide on a valid interpretation. In some problems a solution may blow up in finite time with respect to one norm, yet be continuable as a solution in an appropriately weakened sense; this situation occurs for nonlinear hyperbolic equations, for which spatial derivatives of a globally defined weak solution can blow up in finite time due to shock formation (Lax [17,18]).

The preceding remarks are illustrated in this paper through the discussion of two examples. In Section 2 we consider an initial boundary value problem for a nonlinear elastic body subjected to constant pressures applied to its surface. The boundary conditions are, for example, appropriate to the case of inflation of a hollow shell under a maintained internal pressure. The stored-energy function of the elastic material is assumed to satisfy a special case of the 'concavity inequality' of Knops, Levine and Payne [13]. This assumption is interpreted as a precise statement about the weakness of the material for large strains. The concavity method [13] is adapted to show that for suitable initial conditions and pressures no weak solution can exist for all time, and that the L^2 norm of the solution possesses a lower bound which blows up in finite time. It is possible that for certain materials this result is connected with the onset of rupture. The lack of a suitable existence theory for nonlinear hyperbolic systems unfortunately prevents us from making any definite assertion concerning blow-up of the solution.

In Section 3 a model problem is considered for which an existence and continuation theorem leads to a proof of blow-up for certain solutions. The problem consists of the semilinear wave equation

$$u_{tt} = \Delta u + |u|^{\gamma-1}u, \quad t > 0,\ x \in \Omega \ ,$$

with boundary conditions

$$u\Big|_{\partial\Omega} = 0 \ , \ t > 0 \ ,$$

where Ω is a bounded domain in $\mathbb{R}^n$ and $\gamma > 1$. Blow-up of weak solutions in various norms is established for suitable initial conditions and under stated hypotheses on γ and n. The results improve those in Ball [1].

Other blow-up theorems for semilinear equations have been proved by Ball [1] for the parabolic problem

$$u_t = \Delta u + |u|^{\gamma-1}u, \quad t > 0, \; x \in \Omega \quad ,$$
$$u\big|_{\partial\Omega} = 0, \quad t > 0 \quad ,$$

(see also Weissler [31] for some relevant continuation results), and by Glassey [9,10] for the nonlinear Schrödinger equation

$$iw_t = \Delta w + |w|^{\gamma-1}w \; , \; t > 0, \; x \in \mathbb{R}^n \quad .$$

In both cases strong hypotheses are made concerning the size of γ relative to n.

2. Dynamic behaviour of an elastic body under pressure.

Consider a nonlinear elastic body which occupies the bounded open set $\Omega \subset \mathbb{R}^3$ in a reference configuration. We suppose that the boundary $\partial\Omega$ of Ω is the disjoint union of piecewise smooth closed surfaces $\partial\Omega_r \, (r=1,\ldots,M)$.

In a typical motion the particle occupying the point $\underset{\sim}{x} \in \Omega$ in the reference configuration is displaced to $\underset{\sim}{u}(\underset{\sim}{x},t)$ at time t. (See Figure 2.)

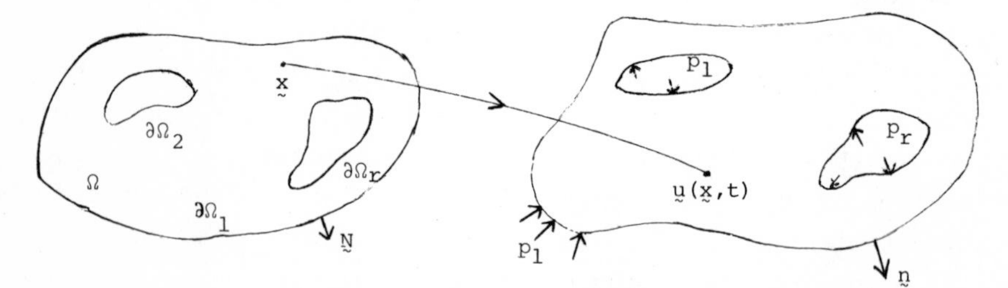

Reference configuration Deformed configuration

Figure 2.

Let $M_+^{3\times3}$ denote the set of real 3×3 matrices with positive determinant. The material properties of the body are characterized by a smooth stored-energy function $W:\Omega \times M_+^{3\times3} \to \mathbb{R}$, in terms of which the total stored energy of the body is given by

$$V = \int_\Omega W(\underset{\sim}{x}, \nabla\underset{\sim}{u}(\underset{\sim}{x},t))\,d\underset{\sim}{x} \quad .$$

Consider the initial boundary value problem

(2.1) $\rho_0(\underset{\sim}{x})\underset{\sim}{u}_{tt} = \operatorname{div} \dfrac{\partial W(\underset{\sim}{x},\nabla\underset{\sim}{u}(\underset{\sim}{x},t))}{\partial \nabla \underset{\sim}{u}}$, $t > 0$, $\underset{\sim}{x} \in \Omega$,

(2.2) $\underset{\sim}{t}(\underset{\sim}{x},t) = -p_r\underset{\sim}{n}(\underset{\sim}{x},t)$, $t > 0$, $\underset{\sim}{x} \in \partial\Omega_r$, $(r=1,\ldots,M)$,

(2.3) $\underset{\sim}{u}(\underset{\sim}{x},0) = \underset{\sim}{u}_0(\underset{\sim}{x})$, $\underset{\sim}{u}_t(\underset{\sim}{x},0) = \underset{\sim}{u}_1(\underset{\sim}{x})$, $\underset{\sim}{x} \in \Omega$,

where $\rho_0 \in L^\infty(\Omega)$ is the density in the reference configuration, $\underset{\sim}{t}$ is the Cauchy stress vector, $\underset{\sim}{n}$ is the unit outward normal to the deformed surface, $p_r(r=1,\ldots,M)$ are constant pressures, and $\underset{\sim}{u}_0, \underset{\sim}{u}_1$ are sufficiently smooth given initial functions. We assume that $\operatorname{ess\,inf}_{\underset{\sim}{x}\in\Omega} \rho_0(\underset{\sim}{x}) > 0$. The hypotheses on $\partial\Omega$ imply the existence of a function $p \in C_0^\infty(\mathbb{R}^3)$ which takes the values p_r on $\partial\Omega_r$.

Let $\tau > 0$, and suppose that $\underset{\sim}{u}$ is a smooth solution of (2.1)- (2.3) on the time interval $[0,\tau]$. Suppose further that $\underset{\sim}{u}(\cdot,t)$ is invertible on $\overline{\Omega}$ with smooth inverse for each $t \in [0,\tau]$. Multiplying (2.1) by a smooth function $\underset{\sim}{v}(\underset{\sim}{x})$ and integrating over Ω we obtain

$$\frac{d}{dt}\int_\Omega \rho_0\underset{\sim}{u}_t\cdot\underset{\sim}{v}\, d\underset{\sim}{x} = \int_{\partial\Omega} \frac{\partial W}{\partial u^i_{,\alpha}} v^i N_\alpha dS - \int_\Omega \frac{\partial W}{\partial u^i_{,\alpha}} v^i_{,\alpha}\, d\underset{\sim}{x} \ ,$$

where $\underset{\sim}{N}$ denotes the unit outward normal to $\partial\Omega$ and where we are using the summation convention for repeated suffices. Applying the definition of $\underset{\sim}{t}$, the boundary conditions (2.2), and the divergence theorem, we obtain

$$\begin{aligned}\int_{\partial\Omega} \frac{\partial W}{\partial u^i_{,\alpha}} v^i N_\alpha dS &= \int_{\partial \underset{\sim}{u}(\Omega)} t_i v^i ds \\ &= - \int_{\partial \underset{\sim}{u}(\Omega)} p v^i n_i ds \\ &= - \int_{\underset{\sim}{u}(\Omega)} \frac{\partial (pv^i)}{\partial u^i} d\underset{\sim}{u} \\ &= - \int_\Omega (pv^i)_{,\alpha} \frac{\partial x^\alpha}{\partial u^i} \det \nabla\underset{\sim}{u}\, d\underset{\sim}{x} \ .\end{aligned}$$

Hence

(2.4) $$\frac{d}{dt}\int_\Omega \rho_0\underset{\sim}{u}_t\cdot\underset{\sim}{v}\, d\underset{\sim}{x} = - \int_\Omega (pv^i)_{,\alpha}(\operatorname{adj}\nabla\underset{\sim}{u})^\alpha_i\, d\underset{\sim}{x} - \int_\Omega \frac{\partial W}{\partial u^i_{,\alpha}} v^i_{,\alpha} d\underset{\sim}{x} \ ,$$

where $\operatorname{adj}\nabla\underset{\sim}{u}$ is the transpose of the matrix of cofactors of $\nabla\underset{\sim}{u}$.

Using the facts that $\partial\Omega_r$ is closed and p_r is constant for each r one can prove (cf Sewell[28], Ball [3]) the following

<u>Transport Lemma</u>

$$\frac{d}{dt} \int_{\partial \underset{\sim}{u}(\Omega)} p\, \underset{\sim}{u} \cdot \underset{\sim}{n}\, ds = 3 \int_{\partial \underset{\sim}{u}(\Omega)} p\, \underset{\sim}{u}_t \cdot \underset{\sim}{n}\, ds \quad .$$

(The coefficient 3 is not a misprint; $\underset{\sim}{u}(\Omega)$ and $\underset{\sim}{n}$ depend on t.)

Multiplying (2.1) by $\underset{\sim}{u}_t$, it follows from the lemma that the energy identity

$$E(t) = E(0), \quad t \in [0,\tau] \quad , \tag{2.5}$$

holds, where

$$E(t) \overset{\text{def}}{=} \int_\Omega [\tfrac{1}{2} \rho_0 |\underset{\sim}{u}_t|^2 + W(\underset{\sim}{x}, \nabla \underset{\sim}{u}) + \tfrac{1}{3} (pu^i)_{,\alpha} (\text{adj} \nabla \underset{\sim}{u})^\alpha_i] d\underset{\sim}{x} \ . \tag{2.6}$$

With the above as motivation we make the following

<u>Definition</u>

Let $D \subset (W^{1,1}(\Omega))^3$. A function $\underset{\sim}{u}:[0,\tau] \to D$ is a <u>weak solution</u> of (2.1) - (2.3) if

(i) $\underset{\sim}{u} \in C^1([0,\tau]; (L^2(\Omega))^3)$ and satisfies (2.3);

(ii) For any $\underset{\sim}{v} \in D$ the integrals on the right-hand side of (2.4) exist and belong to $C([0,\tau])$,

$\int_\Omega \rho_0 \underset{\sim}{u}_t \cdot \underset{\sim}{v}\, d\underset{\sim}{x} \in C^1([0,\tau])$, and (2.4) holds;

(iii) $E(t)$ is well defined for all $t \in [0,\tau]$ and satisfies the energy inequality

$$E(t) \leq E(0), \quad t \in [0,\tau] \quad ;$$

(iv) For each $t \in [0,\tau]$

$$\det \nabla \underset{\sim}{u}(\underset{\sim}{x},t) > 0 \quad \text{for almost all } \underset{\sim}{x} \in \Omega \quad .$$

<u>Remarks</u>: Property (iii) is consistent with the use of entropy conditions in the theory of nonlinear hyperbolic conservation laws; energy may be dissipated by shock waves. Property (iv) could be strengthened by requiring that $\underset{\sim}{u}$ be invertible, but we do not assume this.

We suppose that for each $\underset{\sim}{x} \in \Omega$, $F \in M^{3\times3}_+$, W satisfies

$$3W(\underset{\sim}{x},F) \geq \frac{\partial W}{\partial F^i_\alpha}(\underset{\sim}{x},F) F^i_\alpha \quad .$$

We write this constitutive inequality in the abbreviated form

$$3W(F) \geq \frac{\partial W}{\partial F}(F) \cdot F \quad . \tag{C}$$

Condition (C) is a special case of the 'concavity inequality' of Knops, Levine and Payne [13]. It is also in a certain sense the opposite of a condition studied in Ball [4]. Consider a homogeneous cube of material of side $\frac{1}{\lambda}$. Fix $F \in M_+^{3\times 3}$ and consider a uniform deformation of the cube with deformation gradient $\nabla \underset{\sim}{u} = \lambda F$. The shape and size of the deformed cube is independent of λ. The total stored energy of the deformation is given by

$$g(\lambda) = \frac{W(\lambda F)}{\lambda^3} \quad . \tag{2.7}$$

Thus the property

$$g(\lambda) \to \infty \quad \text{as} \quad \lambda \to \infty \tag{2.8}$$

can be viewed as characterizing a material which is 'strong' for large strains. This was the condition proposed in Ball [4]. If we also suppose that

$$W(F) \to \infty \quad \text{as} \quad \det F \to 0 \tag{2.9}$$

(i.e. that infinite energy is required to effect a compression to zero volume) then clearly $g(\lambda)$ tends to infinity as $\lambda \to 0+$. Hence the graph of g for a strong material has the general form shown in Figure 3(a).

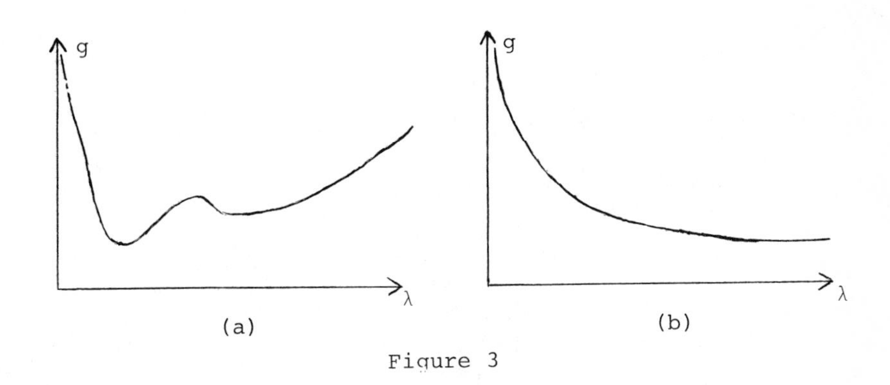

(a) (b)

Figure 3

A condition which might be satisfied by a 'weak' material is that $g'(\lambda) \leq 0$ (see Figure 3b). Differentiating (2.7) it is clear that this condition is equivalent to (C). Condition (C) is satisfied, for example, by stored energy functions of the form

$$W(F) = \text{tr}(FF^T) + h(\det F) \quad ,$$

where $sh'(s) \leq h(s)$ for all $s > 0$; such W can satisfy (2.9).

Theorem 2.1

Let W satisfy condition (C). Let $E(0) < 0$, or $E(0) = 0$ and $\int_\Omega \rho_0 \underset{\sim}{u}_0 \cdot \underset{\sim}{u}_1 d\underset{\sim}{x} > 0$. Let $\underset{\sim}{u}$ be a weak solution of (2.1) - (2.3) on $[0,\tau]$, and define

$$F(t) = \int_\Omega \rho_0 |\underset{\sim}{u}(\underset{\sim}{x},t)|^2 d\underset{\sim}{x} \quad .$$

Then

$$F(t) \geq \frac{a}{(1-kt)^4} \quad ,$$

where a and k are positive constants depending on $\int_\Omega \rho_0 |\underset{\sim}{u}_0|^2 d\underset{\sim}{x}$, $\int_\Omega \rho_0 \underset{\sim}{u}_0 \cdot \underset{\sim}{u}_1 d\underset{\sim}{x}$ and $E(0)$. In particular, if $[0,t_{max})$ is the maximal half-open interval of existence of a weak solution $\underset{\sim}{u}$, then

$$t_{max} \leq k^{-1} < \infty \quad .$$

Proof

Differentiating $F(t)$ twice with respect to t and using (2.4) we obtain

$$\dot{F}(t) = 2 \int_\Omega \rho_0 \underset{\sim}{u} \cdot \underset{\sim}{u}_t d\underset{\sim}{x} \quad , \tag{2.10}$$

$$\ddot{F}(t) = 2 \int_\Omega \rho_0 |\underset{\sim}{u}_t|^2 d\underset{\sim}{x} - 2 \int_\Omega (pu^i)_{,\alpha} (\text{adj}\ \nabla \underset{\sim}{u})^\alpha_i d\underset{\sim}{x} - 2 \int_\Omega \frac{\partial W}{\partial u^i_{,\alpha}} u^i_{,\alpha} d\underset{\sim}{x} \quad . \tag{2.11}$$

Substituting for the second integral in (2.11) from the energy inequality gives

$$\ddot{F}(t) \geq 5 \int_\Omega \rho_0 |\underset{\sim}{u}_t|^2 d\underset{\sim}{x} + 2 \int_\Omega \left[3W(\underset{\sim}{x},\nabla \underset{\sim}{u}) - \frac{\partial W}{\partial u^i_{,\alpha}}(\underset{\sim}{x},\nabla \underset{\sim}{u}) u^i_{,\alpha}\right] d\underset{\sim}{x} - 6E(0) \quad . \tag{2.12}$$

Suppose that $E(0) \leq 0$ and $\int_\Omega \rho_0 \underset{\sim}{u}_0 \cdot \underset{\sim}{u}_1 d\underset{\sim}{x} > 0$. By condition (C) and (2.12),

$$\ddot{F}(t) \geq 5 \int_\Omega \rho_0 |\underset{\sim}{u}_t|^2 d\underset{\sim}{x} \quad .$$

Multiplying by $F(t)$ and using (2.10) and Schwarz's inequality we get

$$F(t)\ddot{F}(t) - \frac{5}{4}\dot{F}^2(t) \geq 0 \quad . \tag{2.13}$$

Let $g(t) = F^{-\frac{1}{4}}(t)$. Then (2.13) becomes

$$\ddot{g}(t) \leq 0 \quad .$$

But $\dot{g}(0) = -\frac{1}{4} F^{-\frac{5}{4}}(0)\dot{F}(0) < 0$. Since

$$g(t) \leq g(0) + \dot{g}(0)t$$

the result follows with

$$a = \int_\Omega \rho_0 |\underset{\sim}{u}_0|^2 d\underset{\sim}{x}, \; k = \int_\Omega \rho_0 \underset{\sim}{u}_0 \cdot \underset{\sim}{u}_1 d\underset{\sim}{x} / 2\int_\Omega \rho_0 |\underset{\sim}{u}_0|^2 d\underset{\sim}{x} \quad .$$

In the case $E(0) < 0$ it follows from (2.12) that $\dot{F}(t_1) > 0$ if $t_1 > \dot{F}(0)/6E(0)$. Then $\dot{g}(t_1) < 0$, $E(t_1) < 0$ so that the previous argument applies. □

Remarks

1. The pressure term in (2.6) may be written in the form

$$\sum_{r=1}^{M} p_r \int_{\underset{\sim}{u}(\partial\Omega_r)} \underset{\sim}{u} \cdot \underset{\sim}{n} \, ds \quad .$$

   Thus if $\int_{\underset{\sim}{u}(\partial\Omega_r)} \underset{\sim}{u}_0 \cdot \underset{\sim}{n} \, ds \neq 0$ for some r it is possible to satisfy the condition $E(0) \leq 0$ by a suitable choice of pressures.

2. Some results in the case $E(0) > 0$ can be obtained using methods of Knops, Levine and Payne [13].

3. If W satisfies

$$3W - \frac{\partial W}{\partial F} \cdot F \geq k$$

   for some constant k, then $W - \frac{k}{3}$ satisfies condition (C), so that Theorem 2.1 may be applied.

4. Some information concerning the stability of equilibrium solutions for certain pressure boundary value problems is contained in Coleman and Dill [5].

3. Blow-up for a semilinear wave equation.

Consider the problem

$$(3.1)\quad \begin{aligned} &u_{tt} = \Delta u + |u|^{\gamma-1}u, \quad t > 0,\ x \in \Omega, \\ &u\big|_{\partial\Omega} = 0,\ t \geq 0;\ u(x,0) = u_0(x),\ u_t(x,0) = u_1(x),\ x \in \Omega, \end{aligned}$$

where Ω is a bounded domain in $\mathbb{R}^n$ with smooth boundary $\partial\Omega$, and where $\gamma > 1$ is a constant satisfying $\gamma \leq \frac{n}{n-2}$ if $n \geq 3$.

Let $X = W_0^{1,2}(\Omega) \times L^2(\Omega)$ and define $A = \begin{pmatrix} 0 & I \\ \Delta & 0 \end{pmatrix}$ with $D(A) = (W_0^{1,2}(\Omega) \cap W^{2,2}(\Omega)) \times W_0^{1,2}(\Omega)$. It is well known that A generates a strongly continuous group $T(\cdot)$ of bounded linear operators on X. Let $f\begin{pmatrix} u \\ v \end{pmatrix} = \begin{pmatrix} 0 \\ |u|^{\gamma-1}u \end{pmatrix}$. We write (3.1) in the form

$$(3.2)\quad \dot{w} = Aw + f(w),\ w(0) = \varphi,$$

where $w = \begin{pmatrix} u \\ u_t \end{pmatrix}$. We assume that $\varphi = \begin{pmatrix} u_0 \\ u_1 \end{pmatrix}$ belongs to X. A weak solution w of (3.1) is by definition a solution of the integral equation

$$(3.3)\quad w(t) = T(t)\varphi + \int_0^t T(t-s)f(w(s))ds \quad .$$

The corresponding function u satisfies (3.1) in the sense of distributions. The hypotheses on γ imply that $f: X \to X$ and is locally Lipschitz. Therefore a standard argument (cf Segal [27]) gives the following existence and continuation theorem.

Proposition 3.1

There exists a unique maximally defined weak solution $w = \begin{pmatrix} u \\ u_t \end{pmatrix}$, $w \in C([0,t_{max});X)$, $t_{max} > 0$, of (3.1). If $t_{max} < \infty$ then

$$\lim_{t \nearrow t_{max}} \|w(t)\|_X = \infty \quad .$$

It can be shown (cf Reed [26], Ball [2]) that the solution u in Proposition 3.1 satisfies the energy equation

$$(3.4)\quad E(u(\cdot,t),u_t(\cdot,t)) = E(u_0,u_1), \qquad t \in [0,t_{max}) \quad ,$$

where $E: X \to \mathbb{R}$ is defined by

$$E(v,y) = \int_\Omega [\tfrac{1}{2}|y|^2 + \tfrac{1}{2}|\nabla v|^2 - \tfrac{1}{\gamma+1}|v|^{\gamma+1}]dx \quad .$$

<u>Notation</u>: $\|\cdot\|_p$ and $\|\cdot\|_{1,p}$ denote the norms in $L^p(\Omega)$ and $W^{1,p}(\Omega)$ respectively. The inner product in $L^2(\Omega)$ is written $(\ ,\ )$.

Global nonexistence results for (3.1) have been proved by Glassey [8], Levine [20] and Tsutsumi [30]. Following the work of these authors we prove

<u>Theorem 3.2.</u>

If $E_0 \overset{def}{=} E(u_0(\cdot),u_1(\cdot)) < 0$, or if $E_0 = 0$ and $(u_0,u_1) > 0$, then $t_{max} < \infty$ and

$$(3.5)\quad \lim_{t\nearrow t_{max}} \|u(t)\|_{\gamma+1} = \lim_{t\nearrow t_{max}} [\|\nabla u(t)\|_2^2 + \|u_t(t)\|_2^2]^{\frac{1}{2}} = \infty \quad .$$

<u>Proof</u>

In view of (3.4) and Proposition 3.1, it suffices to show that $t_{max} < \infty$. Suppose $t_{max} = \infty$. Let $F(t) = \|u(t)\|_2^2$. Using (3.1) and (3.4) we obtain the differential inequality

$$(3.6)\quad \ddot{F}(t) \geq \frac{2(\gamma-1)}{\gamma+1} \int_\Omega |u|^{\gamma+1}dx - 4E_0 \geq kF^{(\gamma+1)/2}(t) - 4E_0 \ ,$$

where $k > 0$ is a constant. Solving this inequality under the given initial conditions leads to a contradiction. For the details the reader is referred to Ball [1]. □

In order to sharpen the blow-up result (3.5) we will make stronger hypotheses on γ.

<u>Theorem 3.3</u>

Let the hypotheses of Theorem 3.2 hold.

(i) Let

$$1 < \gamma \leq 1 + \frac{4}{n} \quad \text{if} \quad n = 2,3 \ ,$$

$$1 < \gamma \leq \frac{n}{n-2} \quad \text{if} \quad n \geq 4 \quad .$$

Then

$$\lim_{t \nearrow t_{max}} (u,u_t)(t) = \infty .$$

(ii) Let $1 \leq p < \gamma + 1$, and

$$1 < \gamma < 1 + \frac{2p}{n} .$$

Suppose also that $\gamma \leq \frac{n}{n-2}$ if $n \geq 3$.

Then

$$\lim_{t \nearrow t_{max}} \|u(t)\|_p = \infty .$$

Proof

(ii) Since $E_0 \leq 0$, (3.4) implies that

$$\int_\Omega |\nabla u|^2 dx \leq \frac{2}{\gamma+1} \int_\Omega |u|^{\gamma+1} dx \tag{3.7}$$

on the interval $[0,t_{max})$.

An interpolation inequality of Gagliardo [7] and Nirenberg [23] (see also Ladyzhenskaya, Solonnikov and Ural'ceva [16]) implies that

$$\|v\|_q \leq K\|v\|_p^{1-a}\|\nabla v\|_2^a , \quad \text{for all} \quad v \in W_0^{1,2}(\Omega) , \tag{3.8}$$

where

$$\frac{1}{q} = (1-a)\frac{1}{p} + a\left(\frac{1}{2} - \frac{1}{n}\right) ,$$

$$\frac{1}{p} \geq \frac{1}{q} \geq \frac{1}{2} - \frac{1}{n} ,$$

and where the constant K depends only on p, q and n.

Applying (3.8) with $q = \gamma + 1$, noting that by hypothesis

$$\frac{1}{\gamma+1} \geq \frac{1}{2} - \frac{1}{n} ,$$

and using (3.7), we obtain

$$\|u\|_{\gamma+1} \leq K\|u\|_p^{1-a}\|\nabla u\|_2^a$$

$$\leq C\|u\|_p^{1-a}\|u\|_{\gamma+1}^{(\gamma+1)a/2} ,$$

where here and below C denotes a generic constant. But

$(\gamma+1)a/2 < 1$ if and only if $\gamma < 1 + \frac{2p}{n}$. The result follows from (3.5).

(i) We first suppose that $n > 1$. Note that if $1 < \gamma < 1 + \frac{4}{n}$ then the result follows from (ii) with $p = 2$, and from the fact that $F(t)$ is convex by (3.6). Thus for $n > 1$ we need only consider the case $\gamma = 1 + \frac{4}{n}$. However, to show how the number $1 + \frac{4}{n}$ arises we shall consider the case of general γ.

Since $\|T(t)\| \leq M$ it follows from (3.3) and Proposition 3.1 that

$$\int_0^{t_{max}} \|f(w(t))\|_X dt = \infty \quad .$$

Thus

$$(3.9) \qquad \int_0^{t_{max}} \left(\int_\Omega |u|^{2\gamma} dx \right)^{\frac{1}{2}} dt = \infty \quad .$$

By (3.6) it suffices to prove that

$$(3.10) \qquad \int_0^{t_{max}} \int_\Omega |u|^{\gamma+1} dx\, dt = \infty \quad .$$

Clearly (3.10) will follow from (3.9) if we show that u satisfies the estimate

$$(3.11) \qquad \|u\|_{2\gamma}^{\gamma} \leq C \|u\|_{\gamma+1}^{b(\gamma+1)}$$

with $0 < b \leq 1$. Applying (3.8) with $q = 2\gamma$ and $p = \gamma+1$, noting that

$$\frac{1}{\gamma+1} > \frac{1}{2\gamma} \geq \frac{1}{2} - \frac{1}{n} \quad ,$$

and using (3.7), we obtain (3.11) with

$$b = \frac{1}{\gamma+1} \left[(1-a)\gamma + a\gamma(\gamma+1)/2 \right] \quad .$$

But $b \leq 1$ if and only if $a \leq \frac{2}{\gamma(\gamma-1)}$, and a short calculation shows that this holds provided

$$\gamma^2 - \frac{4\gamma}{n} - \left(1 + \frac{4}{n}\right) \leq 0 \quad .$$

The result for $n > 1$ follows.

Finally we consider the case $n = 1$. Let $Y = W_0^{1,1}(\Omega) \times L^1(\Omega)$. It is well known that A generates a strongly continuous group $T(t)$ on Y. (This follows immediately from D'Alembert's solution of the wave equation; it is false for $n > 1$ (cf Littman [22]).) It is easily verified that $f:Y \to Y$ and is locally Lipschitz. Hence there exists a unique maximally defined solution $w = \binom{u}{u_t}$, $w \in C([0,t_1);Y), t_1 > 0$, of (3.1). Clearly $t_1 \geq t_{max}$. But by (3.4)

$$\lim_{t \nearrow t_{max}} \|u(t)\|_{\gamma+1} = \infty \quad .$$

Hence

$$\lim_{t \nearrow t_{max}} \|u(t)\|_{1,1} = \infty \quad ,$$

so that $t_1 = t_{max}$.

As before we deduce that

$$(3.12) \qquad \int_0^{t_{max}} \|f(w(t))\|_Y dt = \int_0^{t_{max}} \int_\Omega |u|^\gamma dx\, dt = \infty \quad .$$

It follows immediately that

$$\int_0^{t_{max}} \int_\Omega |u|^{\gamma+1} dx\, dt = \infty \quad ,$$

so that (i) holds. □

Remarks

1. Equations (3.9), (3.10) and (3.12) may be interpreted as statements about rate of blow-up. By integrating the inequality in (3.6) one may also obtain the upper bound

$$\|u(t)\|_2 \leq C(t_{max} - t)^{2/(1-\gamma)} \quad .$$

2. Similar results to Theorems 3.2 and 3.3 can be obtained by exactly the same methods for the problem

$$u_{tt} = \Delta u + f(u)$$

$$u|_{\partial\Omega} = 0 \;;\; u(x,0) = u_0(x),\; u_t(x,0) = u_1(x) \quad ,$$

under suitable polynomial growth hypotheses on f.

3. Little seems to be known about the global behaviour of all solutions of (3.1), although some results have been obtained by Payne and Sattinger [25] based on the study of potential wells. For example, is it true that every solution either blows up in finite time or remains bounded for all time?

REFERENCES

1. J. M. Ball, Remarks on blow-up and nonexistence theorems for nonlinear evolution equations, Quart. J. Math., Oxford 28 (1977) 473-486.
2. J. M. Ball, On the asymptotic behaviour of generalized processes, with applications to nonlinear evolution equations, J. Differential Eqns. 27 (1978) 224-265.
3. J. M. Ball, Convexity conditions and existence theorems in nonlinear elasticity, Arch. Rat. Mech. Anal. 63 (1977) 337-403.
4. J. M. Ball, Constitutive inequalities and existence theorems in nonlinear elastostatics, in 'Nonlinear analysis and mechanics', Vol. 1, ed. R. J. Knops, Pitman, 1977.
5. B. D. Coleman and E. H. Dill, On the stability of certain motions of incompressible materials with memory, Arch. Rat. Mech. Anal. 30 (1968) 197-224.
6. H. Fujita, On the blowing up of solutions of the Cauchy problem for $u_t = \Delta u + u^{1+\alpha}$, J. Fac. Sci. Univ. Tokyo Sect. 1 13 (1966) 109-124.
7. E. Gagliardo, Ulteriori proprietà di alcuno classi di funzioni in piu variabili, Ricerche Mat. 8 (1959) 24-51.
8. R. T. Glassey, Blow-up theorems for nonlinear wave equations, Math. Z. 132 (1973) 183-203.
9. R. T. Glassey, On the blowing-up of solutions to the Cauchy problem for nonlinear Schrödinger equations, J. Math. Phys. 18 (1977) 1794-1797.
10. R. T. Glassey, Personal communication.
11. P. Hartman, "Ordinary differential equations", Wiley, New York, 1964.
12. S. Kaplan, On the growth of solutions of quasilinear parabolic equations, Comm. Pure Appl. Math. 16 (1963) 327-330.

13. R. J. Knops, H. A. Levine and L. E. Payne, Non-existence, instability, and growth theorems for solutions of a class of abstract nonlinear equations with applications to nonlinear elastodynamics, Arch. Rat. Mech. Anal. 55 (1974) 52-72.

14. R. J. Knops and B. Straughan, Non-existence of global solutions to non-linear Cauchy problems arising in mechanics, Symp. Trends in Appl. of Pure Math. to Mechanics, Pitman, 1975, 187-206.

15. V. K. Kalantarov and O. A. Ladyzhenskaya, Blow-up theorems for quasilinear parabolic and hyperbolic equations, (in Russian) Zap. Nauk, Semin. Leningrad Otd. Mat. Inst. Steklov 69 (1977) 77-102.

16. O. A. Ladyzhenskaya, V. A. Solonnikov, and N. N. Ural'ceva, "Linear and quasilinear equations of parabolic type," Amer. Math. Soc., Translations of Mathematical Monographs, Vol. 23, 1968.

17. P. D. Lax, Hyperbolic systems of conservation laws, II, Comm. Pure Appl. Math. 10 (1957) 537-566.

18. P. D. Lax, Development of singularities of solutions of nonlinear hyperbolic partial differential equations, J. Math. Phys. 5 (1964) 611-613.

19. H. A. Levine, Some nonexistence and instability theorems for formally parabolic equations of the form $Pu_t = - Au + F(u)$, Arch. Rat. Mech. Anal. 51 (1973) 371-386.

20. H. A. Levine, Instability and nonexistence of global solutions to nonlinear wave equations of the form $Pu_{tt} = - Au + F(u)$, Trans. Amer. Math. Soc. 192 (1974) 1-21.

21. H. A. Levine, Nonexistence of global weal solutions to some properly and improperly posed problems of mathematical physics: the method of unbounded Fourier coefficients, Math. Ann. 214 (1975) 205-220.

22. W. Littman, The wave operator and L_p norms, J. Math. Mech. 12 (1963) 55-68.

23. L. Nirenberg, On elliptic partial differential equations, Ann. Scuola Norm. Sup. Pisa, 13 (1959) 115-162.

24. L. E. Payne, 'Improperly posed problems in partial differential equations,' SIAM Regional conference series in applied mathematics, Vol. 22, 1975.

25. L. E. Payne and D. H. Sattinger, Saddle points and instability of nonlinear hyperbolic equations, Israel J. Math. 22 (1975) 273-303.

26. M. Reed, "Abstract non-linear wave equations," Springer Lecture notes in Mathematics, Vol. 507, 1976.

27. I. Segal, Non-linear semi-groups, Ann. Math. 78 (1963) 339-364.

28. M. J. Sewell, On configuration-dependent loading, Arch. Rat. Mech. Anal. 23 (1967) 327-351.

29. B. Straughan, Further global nonexistence theorems for abstract nonlinear wave equations, Proc. Amer. Math. Soc. 48 (1975) 381-390.

30. M. Tsutsumi, On solutions of semilinear differential equations in a Hilbert space, Math. Japonicae 17 (1972) 173-193.

31. F. B. Weissler, Semilinear evolution equations in Banach spaces, to appear.

Acknowledgement. I am indebted to M. G. Crandall for his perceptive comments on a previous version of this paper. In particular, he improved the range of values of γ in Theorem 3.3 (ii) and greatly simplified the proof.

Department of Mathematics
Heriot-Watt University
Edinburgh, Scotland

A Hamiltonian Approach to the *K dV* and Other Equations

Peter D. Lax

1. Introduction.

Recent investigations, originating in the fundamental work of Kruskal, Zabusky and Gardner reveal that a large number of equations governing nonlinear wave motion can be put in a Hamiltonian formalism. What is even more astonishing, many of these equations turn out to be completely integrable. In this talk we describe a miniscule generalization of the Hamiltonian formalism, and describe its applications to the KdV equation, known for a long time, see [7] and [4], and to the regularized long wave equation (RLW) championed by Benjamin and others, see [2] and [19].

2. Hamiltonian Mechanics.

In this section we briefly review classical Hamiltonian Mechanics, and put it in a form suitable for infinite dimensional phase space.

The phase space of the classical theory is 2N-dimensional space, or some portion of it; the coordinates in this space are denoted by p_j, q_j, $j=1,\ldots,N$. A Hamiltonian is any sufficiently differentiable function H in phase space; the Hamiltonian form of the equations of motion is

$$\frac{d}{dt}p_j = -H_{q_j}, \quad \frac{d}{dt}q_j = H_{p_j}, \quad j=1,\ldots,N \quad , \tag{2.1}$$

where the subscripts of H indicate partial derivatives.

ISBN 0-12-195250-9

Let F and K be any two functions in phase space; their Poisson bracket is defined to be

$$[F,K] = \sum_{1}^{N}(F_{q_j}K_{p_j} - F_{p_j}K_{q_j}) \quad . \tag{2.2}$$

It is convenient to rewrite both (2.1) and (2.2) in block vector and matrix notation. We write

$$\begin{bmatrix} p \\ q \end{bmatrix} = u, \qquad \begin{bmatrix} H_p \\ H_q \end{bmatrix} = H_u \quad . \tag{2.3}$$

Then we can write (2.1) as

$$\frac{d}{dt}u = J_0 H_u \quad , \tag{2.4}$$

where

$$J_0 = \begin{bmatrix} 0 & -I \\ I & 0 \end{bmatrix} \quad , \tag{2.5}$$

I denoting the $N\times N$ unit matrix, 0 the $N\times N$ zero matrix. The Poisson bracket (2.2) can be written as

$$[F,K] = (F_u, J_0 K_u) \quad , \tag{2.6}$$

where the parentheses (,) are defined by

$$(u,u') = \sum(p_j p_j' + q_j q_j') \quad . \tag{2.7}$$

In terms of (,) the gradient H_u of H can be defined by

$$\frac{d}{d\varepsilon}H(u+\varepsilon w)_{\varepsilon=0} = (H_u, w) \quad . \tag{2.8}$$

We ask (and answer) the following entirely elementary question:

When can a system of equations of the form

$$\frac{d}{dt}v = JK_v \quad , \tag{2.9}$$

J some given constant matrix, K some function in phase space, be reduced to Hamiltonian form by a linear change of variables

$$u = Tv \quad , \tag{2.10}$$

T a constant matrix independent of t? We define the transformed Hamiltonian H by

$$H(u) = H(Tv) = K(v) \quad .$$

Using (2.8) we get

$$T^*H_u = K_v \quad , \tag{2.11}$$

where T^* is the transpose of T. Multiplying (2.9) by T and using (2.10), (2.11) we get

$$\frac{d}{dt} u = TJT^*H_u \quad .$$

This is of Hamiltonian form iff

$$TJT^* = J_0 \quad . \tag{2.12}$$

Given a real matrix J, this relation can be satisfied by some real T iff J satisfies

(2.13)
$$\text{i)} \quad J^* = -J \quad ,$$
$$\text{ii)} \quad J \text{ is nonsingular} \quad .$$

The conditions are obviously necessary; to show their sufficiency we note that by antisymmetry and reality, the eigenvalue of J are purely imaginary; since J is nonsingular the eigenvalues are nonzero, and the eigenvectors come in conjugate pairs:

$$Jf_k = i\lambda_k f_k, \quad \lambda_k > 0, \quad k=1,\ldots,N \quad . \tag{2.14}$$

Writing

$$f = g+ih$$

we can rewrite (2.14) as

$$Jg = -\lambda h, \quad Jh = \lambda g \quad .$$

Clearly, if we set

$$T_1^* = (g_1,\ldots,g_N, h_1,\ldots,h_N) \quad ,$$

then

$$T_1JT_1^* = \begin{bmatrix} 0 & -\Lambda \\ \Lambda & 0 \end{bmatrix}$$

where Λ is the diagonal matrix

$$\Lambda = \begin{bmatrix} \lambda_1 & & \\ & \ddots & \\ & & \lambda_N \end{bmatrix} .$$

Define T_2 by

$$T_2 = \begin{bmatrix} \Lambda^{-\frac{1}{2}} & 0 \\ 0 & \Lambda^{-\frac{1}{2}} \end{bmatrix} .$$

Clearly $T = T_2 T_1$ satisfies (2.12).

Given a real inner product space, a real valued function H defined on that space is called differentiable if the directional derivatives (2.8) exist for all directions w, depend linearly on w, and can be represented in form (2.8), with H_u and element of the space. H is called twice differentiable if the directional derivatives of H_u exist in every direction v, and can be represented in the form

$$\frac{d}{d\varepsilon} H_u(u+\varepsilon v) = H_{uu} v \quad , \tag{2.15}$$

where H_{uu} is a linear operator mapping the space into itself. H_{uu}, which is an operator valued function of u, is called the <u>second</u> <u>derivative</u> of H; H_{uu} is a <u>symmetric</u> <u>operator</u>:

$$(H_{uu} v, w) = (v, H_{uu} w) \quad .$$

We shall call an equation of form

$$u_t = JH_u \tag{2.16$_H$}$$

Hamiltonian if J is an antisymmetric operator, independent of u:

$$J^* = -J \quad . \tag{2.17}$$

Since we have dropped the condition (2.13)ii) requiring nondegeneracy, equation $(2.16)_H$ cannot quite be put into genuine Hamiltonian form; thus even in the finite dimensional case we have a miniscule generalization. The important fact is that equations of form (2.16) share all important properties of Hamiltonian equations, and find applications in the examples presented in Section 3.

Properties of Hamiltonian motion are most conveniently expressed in terms of the <u>Poisson bracket</u>, defined by the analogue of formula (2.6) for any pair of differentiable functions F and K:

$$[F,K] = (F_u, JK_u) \quad . \tag{2.18}$$

This Poisson bracket has the usual properties:

a) $[F,K]$ is a bilinear function of F and K.

b) $[F,K] = -[K,F]$.

c) The Jacobi identity

$$[[F,H],K] + [[H,K],F] + [[K,F],H] = 0 \tag{2.19}$$

holds.

Part a) is an immediate consequence of the definition (2.18), and b) follows from the antisymmetry of J. To prove c) we have to calculate the gradient of $[F,H]$. Using the definition of gradient given in (2.8), the antisymmetry of J, its independence of u, and the symmetry of the second derivatives F_{uu} and H_{uu} we get, after a brief calculation, that

$$[F,H]_u = F_{uu}JH_u - H_{uu}JF_u \quad . \tag{2.20}$$

From this (2.19) is easily deduced, using once more the symmetry of F_{uu}, H_{uu} and K_{uu}.

Suppose that $u(t)$ satisfies (2.16); for any function $F(u)$

$$\frac{d}{dt}F = (F_u, \frac{d}{dt}u) = (F_u, JH_u) = [F,H].$$

It follows from this that F <u>is constant along trajectories of the Hamiltonian flow</u> $(2.16)_H$ iff $[F,H] = 0$.

Since $[F,H] = 0$ is a symmetrical relation between F and H, it also follows from $[F,H] = 0$ that H is constant along the trajectories of the Hamiltonian flow $(2.16)_F$. Furthermore the following classical result holds:

If $[F,H] = 0$, <u>then the Hamiltonian flows</u> $(2.16)_H$ <u>and</u> $(2.16)_F$ <u>commute</u>.

A proof of this proposition can be based on the Jacobi identity, from which one can deduce that the Liouville operators associated with the two Hamiltonian flows commute. Here we give another derivation:

Consider two equations of evolution, not necessarily Hamiltonian, of the form

$$u_t = A(u) \quad , \quad v_t = B(v) \quad . \tag{2.21}$$

Denote the associated solution operators by $S_A(r)$ and $S_B(t)$:

$$u(r) = S_A(r)u(0) \quad , \quad v(t) = S_B(t)v(0) \quad . \tag{2.22}$$

The solution operators form one parameter groups:

$$\begin{aligned} S_A(r_1 + r_2) &= S_A(r_1)S_A(r_2) \\ S_B(t_1 + t_2) &= S_B(t_1)S_B(t_2) \end{aligned} \tag{2.23}$$

It follows from the group property that

$$S_A(r) = S_A^n(r/n), \; S_B(t) = S_B^n(t/n) \quad . \tag{2.23'}$$

Therefore we can write, setting $r/n = \rho$, $t/n = \tau$,

$$\begin{aligned} S_A(r)S_B(t) &= S_A^n(\rho)S_B^n(\tau) = \\ &= \sum_{j,k} S_A^{n-j}(\rho)S_B^{k-1}(\tau)CS_B^{n-k}(\tau)S_A^{j-1}(\rho) + S_B(t)S_A(r) \quad , \end{aligned} \tag{2.24}$$

where C abbreviates the commutator

$$C = S_A(\rho)S_B(\tau) - S_B(\tau)S_A(\rho) \quad . \tag{2.25}$$

The number of terms on the right is n^2; since both ρ and τ are $0(1/n)$, we can by letting $n \to \infty$ in (2.24) deduce

$$S_A(r)S_B(t) = S_B(t)S_A(r) \tag{2.26}$$

provided that

$$C = 0(\rho^3 + \tau^3) \quad . \tag{2.27}$$

It follows from (2.21) that

$$(2.28)\qquad \begin{aligned} S_A(\rho) &= I + \rho A + \frac{\rho^2}{2} A_2 + 0(\rho^3) \\ S_B(\tau) &= I + \tau B + \frac{\tau^2}{2} B_2 + 0(\tau^3) \end{aligned}$$

where

$$A_2(u) = u_{tt} = A_u(u)u_t = A_u(u)A(u)$$

and similarly

$$B_2(u) = B_u(u)B(u) \quad ,$$

where $A_u(u)$, $B_u(u)$ denote the Fréchet derivatives of A, B:

$$(2.29)\qquad \frac{d}{d\varepsilon}A(u+\varepsilon v)\Big|_{\varepsilon=0} = A_u(u)v \quad , \text{ etc.}$$

Using (2.28), and (2.29), we get for any vector g, that modulo $0(\rho^3 + \tau^3)$

$$S_A(\rho)S_B(\tau)g = (I + \rho A + \frac{\rho^2}{2} A_2)(g + \tau B(g) + \frac{\tau^2}{2} B_2(g))$$

$$= g + \tau B(g) + \frac{\tau^2}{2} B_2(g) + \rho A(g) + \rho\tau A_u(g)B(g) + \frac{\rho^2}{2} A_2(g) \quad .$$

Similarly

$$S_B(\tau)S_A(\rho)g = g + \rho A(g) + \frac{\rho^2}{2} A_2(g) + $$
$$+ \tau B(g) + \tau\rho B_u(g)A(g) + \frac{\tau^2}{2} B_2(g) \quad .$$

Clearly

$$Cg = S_A(\rho)S_B(\tau)g - S_B(\tau)S_A(\rho)g = 0(\rho^3 + \tau^3)$$

for all g iff

$$(2.30)\qquad A_u(g)B(g) - B_u(g)A(g) = 0 \quad .$$

So if (2.30) holds, the solution operators of (2.21) commute.

Take the Hamiltonian case $A = JF_u$, $B = JH_u$; then

$$(2.31)\qquad A_u(u) = JF_{uu},\ B_u(u) = JH_{uu} \quad .$$

Suppose that

$$[F,H] = (F_u, JH_u) = 0 \tag{2.32}$$

for all u. Putting $u + \varepsilon g$ in place of u and differentiating with respect to ε we get

$$(F_{uu}g, JH_u) + (F_u, JH_{uu}g) = 0 \quad .$$

Using the symmetry of F_{uu}, H_{uu} and the antisymmetry of J we get

$$(g, F_{uu}JH_u - H_{uu}JF_u) = 0$$

for all g; therefore

$$F_{uu}JH_u - H_{uu}JF_u = 0 \quad .$$

Multiplying this by J and using (2.31) we deduce (2.30). This proves that if (2.32) holds, the Hamiltonian flows commute.

We refer to the interesting paper [18] of Olver for a different treatment of the commutation of flows defined by evolution equations.

We turn now to the concept of a completely integrable Hamiltonian system:

A Hamiltonian system $(2.16)_H$ is called completely integrable if there exist Hamiltonians $H_1,\ldots,H_N$, of which the first is the given one, $H_1 = H$, such that

a) $[H_i, H_j] = 0$ for $i, j=1,\ldots N$

b) the H_i are independent, in the sense that their gradients are linearly independent except on a singular set of dimension $< 2N$.

Let u_0 be a nonsingular point; the Hamiltonian trajectory through u_0 lies on the N-dimensional manifold satisfying $H_i(u) = H_i(u_0)$, $i=1,\ldots N$. This manifold can also be generated by the N trajectories $S_{H_j}(t)$ starting at u_0. Since the solution operators $S_{H_j}(t_j)$ commute, it is not hard to prove that the manifold is the product of N lines or circles. In particular, if the manifold is compact, it is a product of N circles.

3. Infinite dimensional Hamiltonian systems.

In this section phase space consists of all real valued C^∞ functions which are periodic, say with period 1; another equally interesting example are the functions of Schwartz class $\mathcal{S}$ on the entire real line $\mathbb{R}$. The inner product is taken to be

$$(u,v) = \int u(x)v(x)dx \quad , \tag{3.1}$$

where the integration is over a single period in the first instance, and all of $\mathbb{R}$ in the second.

Below we list some functions of u and their gradients defined in this phase space:

$$F_1(u) = \int \tfrac{1}{2} u^2 dx, \qquad F_{1_u} = u \quad . \tag{3.2$_1$}$$

$$F_2(u) = \int (\tfrac{1}{6} u^3 - \tfrac{1}{2} u_x^2) dx, \qquad F_{2_u} = \tfrac{1}{2} u^2 + u_{xx} \quad . \tag{3.2$_2$}$$

To build Hamiltonian systems we need to choose some operator J. The choice

$$J = D \tag{3.3}$$

was suggested by Clifford Gardner. Note that this J is anitsymmetric and independent of u, but singular.

For this choice of J, the Hamiltonian equations (2.16) associated with the Hamiltonians F_1 and F_2 given in $(3.2)_1$ and $(3.2)_2$, respectively, are

$$u_t = JF_{1_u} = Du = u_x \tag{3.4$_1$}$$

$$u_t = JF_{2_u} = D(\tfrac{1}{2} u^2 + u_{xx}) = uu_x + u_{xxx} \quad . \tag{3.4$_2$}$$

Equation $(3.4)_1$ is linear and easily solved: $u = u(x+t)$. Equation $(3.4)_2$ on the other hand is nonlinear and far from being easy to solve; it is the celebrated Korteweg-de Vries (KdV) equation.

Clearly, since F_2 is independent of x, F_2 is translation invariant; i.e. F_2 is constant along trajectories of $(3.4)_1$. This is the case iff $[F_1, F_2] = 0$; of course this relation can be verified explicitely. We show now, following

Gardner, Kruskal, Miura and Zabusky, that F_1 and F_2 are merely the first two terms of an infinite sequence of functionals F_i such that

(3.5) $$[F_i, F_j] = 0 \quad .$$

A recursion relation for the gradients $G_j = F_{j_u}$ has been given by Lenard, see [13], in terms of an auxiliary operator L:

(3.6) $$L = D^3 + \frac{2}{3} D + \frac{1}{3} u_x, \; D = \frac{d}{dx} \quad .$$

This operator is antisymmetric:

(3.6)' $$L^* = -L \quad .$$

The Lenard recursion is

(3.7) $$LG_k = JG_{k+1} \quad ,$$

J given by (3.3). It is easy to verify that $G_1 = u$ and $G_2 = \frac{1}{2} u^2 + u$ satisfy this relation. We solve recursively for G_k, $k = 3,4,\ldots$; here one has to verify that the left side satisfies the compatibility relation $\int LG_k dx = 0$. More important, one has to verify that the resulting function G_{k+1} is a gradient. According to the Poincaré lemma this is so iff the Fréchet derivative of G_{k+1} is a symmetric operator; i.e., define N_{k+1} by

(3.8) $$\frac{d}{d\varepsilon} G_{k+1}(u+\varepsilon v)\big|_{\varepsilon=0} = N_{k+1}(u)v \quad ,$$

we have show that N_{k+1} is symmetric. Differentiate (3.7) in the direction v; we obtain

(3.9) $$L' G_k + LG'_k = JG'_{k+1} \quad .$$

Now define the operator M_k by

(3.10) $$L' G_k = \frac{2}{3} vG_{kx} + \frac{1}{3} v_x G_k = M_k v \quad .$$

Substituting (3.10) and (3.8) into (3.9) we get

$(3.11)_k$ $$M_k + LN_k = JN_{k+1} \quad .$$

Similarly

$$(3.11)_{k-1} \qquad M_{k-1} + LN_{k-1} = JN_k \quad .$$

Multiplying $(3.11)_k$ by J on the right, $(3.11)_{k-1}$ by L on the right and subtracting we get

$$(3.12) \qquad JN_{k+1}J = LN_kJ + JN_kL - LN_{k-1}L$$

$$+ M_kJ - M_{k-1}L \quad .$$

Assume that N_k, N_{k-1} are symmetric; then it follows from (3.12) that so is $JN_{k+1}J$ provided that

$$M_kJ - M_{k-1}L$$

is symmetric. This can be verified with a calculation which shows the following: Define the operator $M = M(u;z)$ by

$$(3.13) \qquad M(u;z)v = \frac{d}{d\varepsilon} L(u+\varepsilon v)z\Big|_{\varepsilon=0} \quad .$$

Suppose that z and w satisfy

$$(3.14) \qquad L(u)z = Jw \quad .$$

Then

$$(3.15) \qquad M(u;z)L - M(u;w)J$$

is symmetric.

From the symmetry of $JN_{k+1}J$ that of N_{k+1} is easily deduced; this completes the inductive proof of the existence of an infinite system of conserved quantities F_k.

For another approach to the Lenard recursion see [18]. Relation (3.5) is easily verified using (3.7) plus the anti-symmetry of J and L: with $i < j$ we have

$$[F_i, F_j] = (G_i, JG_j) =$$

$$= (G_i, LG_{j-1}) = -(LG_i, G_{j-1})$$

$$= -(JG_{i+1}, G_{j-1}) = (G_{i+1}, JG_{j-1}) = \dots$$

$$= (G_{i+k-1}, LG_{j-k})$$

$$= (G_{i+k}, JG_{j-k}) \quad .$$

Choosing $k = \frac{j-i+1}{2}$ or $k = \frac{j-i}{2}$, depending on the parity of $j-i$; since both L and J are antisymmetric, we conclude that $[F_i, F_j] = 0$ in either case. In particular $[F_2, F_j] = 0$, so that each F_j is constant along the trajectories of $(3.4)_2$, i.e. the KdV equation has infinitely many conserved functionals. We show now that this is related to the existence of so-called solitons.

A single soliton is merely a traveling wave solution of the KdV equation $(3.4)_2$, i.e. a solution of the form

$$u(x,t) = s(x-ct) \quad , \tag{3.16}$$

which dies down as $x \to \pm \infty$. Setting (3.16) into $(3.4)_2$ gives

$$-cs' = ss' + s''' \quad ; \tag{3.17}$$

for $c > 0$ this equation has a solution $s(x;c)$, unique up to translation, which tends to 0 as $x \to \pm \infty$; $s(x,c)$ can be expressed as a hyperbolic secant. The remarkable discovery of Kruskal and Zabusky was that every solution $u(x,t)$ of KdV has a number of solitons hidden in it, propagating with speeds $c_1,\ldots,c_N$; these solitons appear as t nears $\pm \infty$, and x nears $\pm \infty$:

$$u(x,t) \sim \sum_1^N s(x - c_j t - \theta_j^\pm, c_j) \tag{3.18}$$

in the sense that the difference of the two sides of (3.18) over any interval of the form $|x - c_j t| < K$, $j=1,\ldots N$, K arbitrary, tends to zero as $t \to \pm \infty$. The c_j are called characteristic speeds of the solution u, the $\theta_j^\pm$ the phase shifts. Clearly a translation of $u(x,t)$ by amount τ in time leaves the characteristic speeds unchanged and adds a constant $c_j\tau$ to the phase shifts. This shows that the characteristic speeds c_j and the differences $\theta_j^+ - \theta_j^-$ are conserved functionals. Gardner, Kruskal and Miura have identified these functionals; they have shown that $c_j = 4\lambda_j$, where λ_j is the jth eigenvalue of the Schrödinger operator

$$S = D^2 + \frac{1}{6} u \quad . \tag{3.19}$$

That the eigenvalues $\lambda(u)$ of S are conserved under the KdV flow is particularly simple to show in the Hamiltonian formalism. We calculate the gradient of λ by standard perturbation calculation: In the eigenvalue equation

$$Sw = \lambda w \tag{3.19'}$$

replace u by $u + \varepsilon v$ and differentiate with respect to ε. We get

$$S\dot{w} + \frac{1}{6} vw = \lambda\dot{w} + \dot{\lambda}w \quad ,$$

where the dot denotes ε differentiation. Taking scalar product with w eliminates w and yields, for w of unit norm,

$$\dot{\lambda} = \frac{1}{6} \int vw^2 dx = (\frac{w^2}{6}, v) \quad .$$

Recalling the definition (2.8) of gradient this shows that

$$\lambda_u = \frac{1}{6} w^2 \quad . \tag{3.20}$$

We form now the Poisson bracket:

$$\begin{aligned} [F_2, \lambda] &= (G_2, J\lambda_u) = \\ &= \frac{1}{6} \int (\frac{1}{2} u^2 + u_{xx})(Dw^2)dx \\ &= \frac{1}{6} \int (u^2 ww_x + 2u_{xx}ww_x)dx \quad . \end{aligned} \tag{3.21}$$

We integrate by parts twice to eliminate u_{xx} and u_x, and we replace the second derivatives of w that appear by using the eigenvalue equation (3.19)'. We get this way

$$\frac{4\lambda}{3} \int uww_x dx \quad . \tag{3.21'}$$

Again we use (3.19)' ; multiplying it by w_x we get

$$\frac{1}{6} uww_x = \lambda ww_x - w_x w_{xx} \quad .$$

Substituting this into (3.21)' we can carry out the integration, since w is a square integrable eigenfunction w and w_x tend to 0 as $x \to \pm \infty$, and we conclude that (3.21)' is 0. This proves that $[F_2, \lambda] = 0$, and so λ is preserved. To identify λ with $c/4$ we refer to [11], page 477 for a simple demonstration.

Already Kruskal et al. have observed that the conserved functionals λ_j are not independent of the F_j constructed earlier but are related to them through their asymptotic development for large j. A particularly elegant version is contained in [15].

We remark that two distinct eigenvalues satisfy $[\lambda,\mu] = 0$.

4.

In this section we investigate Hamiltonian systems associated with another choice for J than (3.3):

$$J = D(1 - D^2)^{-1} \quad . \tag{4.1}$$

Note that J is antisymmetric and independent of u. We take

$$(4.2)_1 \qquad H_1(u) = \int (\tfrac{1}{2} u^2 + \tfrac{1}{2} u_x^2)\,dx \quad ,$$

then

$$(4.3)_1 \qquad H_{1_u} = u - u_{xx} \quad .$$

The Hamilton equation (2.16) is

$$(4.4)_1 \qquad u_t = JH_{1_u} = D(1 - D^2)^{-1}(u - u_{xx}) = u_x \quad .$$

Equation $(4.3)_1$ is identical with $(3.4)_1$ and its solution is $u = u(x + t)$. Next we choose

$$(4.2)_2 \qquad H_2(u) = \int (\tfrac{1}{6} u^3 + \tfrac{1}{2} u^2)\,dx \quad ,$$

then

$$(4.3)_2 \qquad H_{2_u} = \tfrac{1}{2} u^2 + u \quad .$$

The Hamilton equation (2.16) is

$$u_t = JH_{2_u} = D(1 - D^2)^{-1}(\tfrac{1}{2} u + u) \quad .$$

Multiplying by $(1 - D^2)$ we get

$$(1 - D^2)u_t = u_t - u_{txx} =$$

$$= D(\frac{1}{2} u^2 + u) = uu_x + u_x \quad ,$$

i.e.

$$u_t = u_x + uu_x + u_{txx} \quad . \tag{4.4$_2$}$$

This is the much-studied regularized long wave equation (RLW), see [2], a model for surface waves in water flowing in a channel, which differs from the KdV equation mainly by the replacement of u_{xxx} by u_{xxt}. Clearly $(4.4)_2$ has $H_1(u)$ as conserved quantity; the question arises, are H_1 and H_2 the first two functionals of a sequence H_k which satisfies $[H_j;H_k] = 0$? So far none have been found (except the nearly trivial $H_0(u) = \int u$). In [19] Olver shows that there are no more whose gradient is a differential operator.

We have seen in the last section that soliton formation is related to the existence of conserved quantities. Equation $(4.4)_2$ has single solitons, i.e. solutions of form (3.16); these are identical in form with those for the KdV equation, except that they are differently parametrized.

Recently Eilbeck [3] has undertaken to study the possible existence of N-soliton solutions by means of numerical calculations. He solved numerically the initial value problem for $(4.4)_2$ starting with an initial configuration

$$u(x,0) = s_1(x) + s_2(x-\ell) \quad ,$$

where s_1 and s_2 are solutions that would propagate with speed c_1 and c_2, respectively, with $c_1 > c_2$. The initial separation ℓ is taken to be so large that the effective overlap between the two solitons was negligible; this is easy to achieve since solitons tend to zero as $x \to \infty$ at an exponential rate. After the lapse of sufficient time T, the numerically computed solution was found to have, within very narrow margin of error, the form

$$u(x,T) \simeq s_1(x-c_1T-\theta_1) + s_2(x-c_2T-\theta_2) \quad .$$

If the difference between the two sides were to tend to zero as ℓ and $T \to \infty$, this would prove the existence of 2-soliton solutions, which is strong evidence for the complete integrability of $(4.4)_2$. However calculations performed in the Soviet Union, [1], indicate a small deviation, about .3%, which persists, no matter how large T is taken.

Recent numerical experiments by Bona and Smith, reported in [19], confirms the persistence of a small deviation, i.e. denies exact soliton formation. The approximate soliton formation observed by Eilbeck et al. may be a property of Hamiltonian systems which are not themselves completely integrable but are perturbations of completely integrable ones. In fact the finite dimensional discretization of the KdV equation on which the numerical experiments are carried out is merely a perturbation of the completely integrable infinite dimensional system; that this perturbation mirrors properties of the completely integrable system is itself a remarkable confirmation of the stability of these properties.

REFERENCES

1. Kh. O. Abdulloev, One More Example of Inelastic Soliton Interaction, Physics Letters, vol. 56A, no. 6, p. 427, May, 1976.
2. T. B. Benjamin, Lectures on nonlinear wave motion, Lectures in Applied Math., vol. 15, American Mathematical Society, Providence, R. I., 1974.
3. J. C. Eilbeck, Numerical Study of the Regularized Long-Wave Equation 1: Numerical Methods, J. Comp. Phys., vol. 19, no. 1, pp. 43-57, Sept., 1975.
4. L. Faddeev and V. E. Zakharov, Korteweg-de Vries equation as completely integrable Hamiltonian system, Funk. Anal. Priloz. 5, pp. 18-27 (1971) (in Russian).
5. H. Flaschka, Integrability of the Toda lattice, Phys. Rev. B, 703 (1974).
6. H. Flaschka, On the Toda lattice, II. Inverse scattering solution, Phys. Rev. B9, 1924 (1974).
7. C. S. Gardner, Korteweg-de Vries equation and generalizations, IV. The Korteweg-de Vries equation as a Hamiltonian system, J. Math. Phys. 12, 1548-1551 (1971).

8. C. S. Gardner, J. M. Greene, M. D. Kruskal and J. M. Miura, Method for solving the Korteweg-de Vries equation, Phys. Rev. Lett. 19, 1095-1097 (1967).

9. C. S. Gardner, M. D. Kruskal and R. M. Miura, Korteweg-de Vries equation and generalizations, II. Existence of conservation laws and constants of motion, J. Math. Phys., 9, 1204-1209 (1968).

10. M. D. Kruskal and N. J. Zabusky, Interaction of "solitons" in a collisionless plasma and the recurrence of initial states, Phys. Rev. Letters, 15, 240-243 (1965).

11. P. D. Lax, Integrals of nonlinear equations of evolution and solitary waves, Comm. Pure Appl. Math. 21, 467-490 (1968).

12. P. D. Lax, Periodic solutions of the KdV equation, Comm. Pure Appl. Math. 28 (1975).

13. P. D. Lax, Almost Periodic Solutions of the KdV Equation, SIAM Review, vol. 18, no. 3, p. 351, July 1976.

14. P. D. Lax, A Hamiltonian approach to the KdV and other equations, Group Theoretical Methods in Physics, Proc. 5th Int. Coll., R. T. Sharp, B. Kolman ed., Acad. Press, New York, pp. 39-57, 1977.

15. H. McKean and P. Van Moerbeke, The spectrum of Hill's equation, Invenciones Mat. 30, 217-274 (1975).

16. H. P. McKean and E. Trubowitz, Hill's operator and hyper-elliptic function theory in the presence of infinitely many branch points, CPAM, vol. XXIX, no. 2, p. 143, March 1976.

17. J. Moser, Three integrable Hamiltonian systems connected with isospectral deformations, Advances in Math., vol. 16, no. 2, p. 197, May 1975.

18. P. J. Olver, Evolution equations possessing infinitely many symmetries, J. Math. Phys., Vol. 18, pp. 1212-1215, 1977.

19. P. J. Olver, Euler operators and conservation laws of the BBM equation, preprint.

20. S. P. Novikov, The periodic problem for the Korteweg-de Vries equation, I. Funk. Anal. Priloz, 8, no. 3, 54-66 (1974) (in Russian).

21. V. E. Zakharov and A. B. Shabat, Soviet Phys. JETP 34, 62 (1972).

Courant Institute of Mathematical Sciences
New York University
New York, New York 10012

A Variational Method for Finding Periodic Solutions of Differential Equations

Paul H. Rabinowitz

§1. Introduction

Our goal here is to describe a method for finding periodic solutions of ordinary and partial differential equations. More accurately it is a procedure for finding critical points of indefinite functionals. Rather than give an abstract formulation of this method, we prefer to illustrate it in a more concrete setting. Accordingly some applications will be stated followed by their detailed treatment by means of our procedure.

We will mainly stay in the setting of Hamiltonian systems of ordinary differential equations. Thus consider such a system:

$$\dot{p} = - H_q \quad , \quad \dot{q} = H_p \tag{1.1}$$

where $p, q \in \mathbb{R}^n$, $H = H(p,q): \mathbb{R}^{2n} \to \mathbb{R}$, and $\cdot$ denotes d/dt. Equivalently (1.1) can be written as

$$\dot{z} = \mathcal{J} H_z \tag{1.2}$$

where $z = (p,q) \in \mathbb{R}^{2n}$ and $\mathcal{J} = \begin{pmatrix} 0 & -I \\ I & 0 \end{pmatrix}$, I denoting the identity matrix in $\mathbb{R}^n$.

Our first result concerns the existence of periodic solutions of (1.2) on a prescribed energy surface:

ISBN 0-12-195250-9

Theorem 1.3: If $H \in C^1(\mathbb{R}^{2n}, \mathbb{R})$ and satisfies

(H_1) $H_z \neq 0$ on $H^{-1}(1)$,

(H_2) $H^{-1}(1)$ is radially diffeomorphic to S^{2n-1}, i.e. the mapping $z \to \frac{z}{|z|}$, $H^{-1}(1) \to S^{2n-1}$ is a diffeomorphism,

then (1.2) possesses a periodic solution on $H^{-1}(1)$.

Observe that the period of this solution is a priori unknown and indeed determining it is one of the main difficulties to be overcome in the course of the proof of Theorem 1.3. An interesting open question under the hypotheses of Theorem 1.3 is whether better lower bounds for the number of geometrically distinct solutions can be given. For the special case of $H(z)$ a positive definite quadratic form plus higher order terms, it has been shown by Weinstein [1] that for each small b, (1.2) has at least n distinct periodic orbits on $H^{-1}(b)$. It is tempting to conjecture that the same lower bound holds for our set-up.

Next we state a result for (1.2) where the period rather than the energy is prescribed.

Theorem 1.4: Suppose $H \in C^1(\mathbb{R}^{2n}, \mathbb{R})$ and satisfies

(H_3) $H(z) \geq 0$ for all $z \in \mathbb{R}^{2n}$,

(H_4) $H(z) = o(|z|^2)$ at $z = 0$,

(H_5) There is an $\bar{r} > 0$ and $\theta \in (0, \frac{1}{2})$ such that $0 < H(z) \leq \theta(z, H_z(z))_{\mathbb{R}^{2n}}$ for $|z| \geq \bar{r}$.

Then for any $\tau > 0$, (1.2) possesses a nonconstant τ periodic solution.

At first glance, Theorems 1.3 and 1.4 appear to be rather different results, but in fact Theorem 1.3 can be obtained as a simple consequence of Theorem 1.4. Alternatively, a direct proof can be given following the lines of our solution procedure. The ideas that are used in the proof of Theorem 1.4 work equally well if H depends explicitly on t in a time periodic fashion, i.e. we have a forced rather than free vibration situation, and one seeks a solution of (1.2) having the same period as the forcing term.

We suspect that a sharper conclusion obtains under the hypotheses of Theorem 1.4, namely for all $\tau > 0$, (1.2) possesses a nonconstant periodic solution with τ as minimal period. To merely get a τ periodic solution does not require the full strength of the hypotheses of Theorem 1.4. In fact we have the following generalization of this result:

Theorem 1.5: Suppose $H \in C^1(\mathbb{R}^{2n}, \mathbb{R})$ and satisfies (H_5). Then for any $\tau, \hat{r} > 0$, there is a τ periodic solution $z(t)$ of (1.2) having $\|z\|_{L^\infty} > \hat{r}$.

Simple examples show the period τ need not be minimal if we only assume (H_5). Theorem 1.4 is of course a consequence of Theorem 1.5. However we prefer to give separate proofs of these results since the latter requires the introduction of some additional topological machinery which can be bypassed in proving Theorem 1.4 directly.

For comparison purposes, we conclude our list of theorems by stating an analogue of Theorem 1.4 for a partial differential equation. Consider the semilinear wave equation

$$(1.6) \quad \begin{cases} u_{tt} - u_{xx} + f(u) = 0 \quad , \quad 0 < x < \pi, \ t \in \mathbb{R} \\ u(0,t) = 0 = u(\pi,t) \ . \end{cases}$$

Theorem 1.7: Suppose $f \in C^2(\mathbb{R}, \mathbb{R})$ and satisfies

(f_1) f is strictly monotone increasing,

(f_2) $f(r) = o(|r|)$ at $r = 0$,

(f_3) there are constants $\bar{r} > 0$ and $\theta \in (0, \frac{1}{2})$ such that $F(r) = \int_0^r f(s)ds \leq \theta\, r\, f(r)$ for $r \geq \bar{r}$.

Then for any τ which is a rational multiple of π, (1.6) possesses a nontrivial classical solution which is τ periodic in t.

The greater technicalities involved in working with (1.6) required imposing more restrictions on the nonlinearity f and on the period τ than in Theorem 1.4. We do not know whether (f_1) or the rationality condition on $\tau\pi^{-1}$ can be eliminated. Likewise it is not known if there is an analogue of Theorem 1.3 in this setting. The details of the proof of Theorem 1.7 can be found in [2] and will not be further discussed here.

Our approach towards the above results is by means of the calculus of variations. We try to find solutions of (1.2) or (1.6) as critical points of corresponding functionals. For example, in the context of Theorem 1.4 with $\tau = 2\pi$, we seek critical points of the corresponding Lagrangian:

$$\int_0^{2\pi} [(p,\dot{q})_{\mathbb{R}^n} - H(z)]dt \tag{1.8}$$

while for (1.6) (and $\tau = 2\pi$) the analogue of (1.8) is

$$\int_0^{2\pi}\int_0^{\pi} [\frac{1}{2}(u_t^2 - u_x^2) - F(u)]dx\,dt \quad . \tag{1.9}$$

To treat the set up of Theorem 1.3, we first make a change of time variable $t \to 2\pi\,\tau^{-1}t \equiv \lambda^{-1}t$, where τ is the unknown period, so that (1.2) transforms to

$$\dot{z} = \lambda\,\mathcal{J}H_z \tag{1.10}$$

and the unknown period becomes 2π. Then working in the class of 2π periodic functions, we search for critical points of the action integral

$$A(z) = \int_0^{2\pi} (p,\dot{q})_{\mathbb{R}^n}\,dt \tag{1.11}$$

subject to the constraint

$$\frac{1}{2\pi}\int_0^{2\pi} H(z)dt = 1 \quad . \tag{1.12}$$

Formally the unknown period then appears in (1.10) via the Lagrange multiplier λ.

As was mentioned at the beginning of this section, the above functionals are indefinite. In particular, they are neither bounded from above nor from below and the quadratic parts of (1.8) and (1.9) have infinite dimensional subspaces on which they are positive and on which they are negative. Thus obtaining critical points of (1.8), (1.9), or (1.11) - (1.12) is a subtle matter and we do not know how to carry this out in any direct fashion. An approximation procedure is used instead. First the functional is restricted to a finite dimensional subspace of $(L^2(S^1))^{2n}$. Secondly a minimax argument is employed to obtain a critical value and corresponding nontrivial critical point for the finite dimensional problem. Thirdly the minimax characterization of the critical value is used to obtain bounds for the critical value and critical point. Having sufficient estimates, we can use standard arguments to pass to a limit to find a solution of (1.2) (or (1.6)). Lastly in the context of Theorems 1.4, 1.5, or 1.7, an additional argument is required to be sure that the solution obtained is nontrivial. We shall give a detailed illustration of this method in §2.

There does not seem to have been much work of the nature of the above theorems in the literature. Our results, Theorems 1.3 - 1.4 can be found in [3]. Theorem 1.5 is new. Earlier Seifert [4] studied the Euler-Lagrange equations corresponding to the Lagrangian $Q - U$ where $Q(x,\dot{x}) = \sum a_{ij}(x)\dot{x}_i\dot{x}_j$ is positive definite in $\dot{x}$, $a_{ij}(x)$ and $U(x)$ are real analytic in a domain $G \subset \mathbb{R}^n$, $U = E$ and $U_x \neq 0$ on ∂G, $U < E$ in G, and G is homeomorphic to the unit ball in $\mathbb{R}^n$. Using geodesic arguments from differential geometry, he showed that the Euler-Lagrange equations for $Q - U$ possess a time periodic solution with energy E. More recently, in work done concurrently with our own, Weinstein [5] extended Seifert's arguments and results replacing $Q - U$ by $H(p,q) = K(p,q) + U(q)$ where U is as above and K is even and convex in p for fixed q. As an application, he obtained a variant of Theorem 1.3 with (H_2) replaced by the condition that $H^{-1}(1)$ bounds a convex region. Some other results of a special nature have been obtained for related problems by Berger [6], Gordon [7], Clark [8], Jacobowitz [9], and Hartman [10]. A considerable amount of work has also been carried out on bifurcation questions for Hamiltonian systems. We refer the reader to Berger [6], Weinstein [1], [11], Moser [12], Bottkol [13], Chow-Mallet-Paret [14], and Fadell-Rabinowitz [15] for more information.

Theorem 1.4 will be proved in §2 using the procedure outlined above. Then an elementary proof of Theorem 1.3 will be carried out in §3 using Theorem 1.4. Lastly in §4 we prove Theorem 1.5. To carry out our method here, we introduce a topological index theory which was developed in [15] and which forms the basis for the minimax constructions used for this theorem.

§2. Proof of Theorem 1.4

The proof of Theorem 1.4 will be given in this section. Observe that no upper bound is placed on the rate of growth of H at infinity. This creates some technical problems which we get around by introducing a new Hamiltonian H_K which coincides with H for $|z| \leq K$ and grows at a prescribed rate at infinity. Let $K > \bar{r}$ and $\chi \in C^\infty(\mathbb{R}^+, \mathbb{R}^+)$ such that $\chi(s) = 1$ if $s \leq K$, $\chi(s) = 0$ if $s \geq K+1$, and $\chi'(s) < 0$ if $s \in (K, K+1)$. Now set

$$H_K(z) = \chi(|z|)H(z) + (1 - \chi(|z|))\rho|z|^4 \tag{2.1}$$

where $\rho = \rho(K)$ satisfies

$$\rho \geq (K+1)^{-4} \max_{|z|=K+1} H(z) . \tag{2.2}$$

Then $H_K \in C^1(\mathbb{R}^{2n}, \mathbb{R})$ and satisfies (H_3) - (H_4). Moreover a calculation using (2.2) and (H_5) shows H_K satisfies (H_5) with θ replaced by $\hat{\theta} = \max(\theta, \frac{1}{4})$. Setting $z = r\,w$ where $w \in S^{2n-1}$, (H_5) implies that

$$\frac{d\,H_K(r\,w)}{dr} \geq \hat{\theta}\, r\, H_K(r\,w) \tag{2.3}$$

for $r > \bar{r}$. On integration (2.3) gives

$$H_K(z) \geq a_1|z|^{\hat{\theta}^{-1}} - a_2 \tag{2.4}$$

for all $z \in \mathbb{R}^{2n}$ where the positive constants a_1, a_2 are independent of K.

The Hamiltonian system corresponding to H_K is

$$\dot{z} = \mathcal{J} H_{Kz} . \tag{2.5}$$

Instead of seeking τ-periodic solutions of (1.2) or (2.5), it is convenient to make the change of variables $t \to 2\pi\,\tau^{-1}\,t \equiv \lambda^{-1}\,t$ transforming (1.2) and (2.5) into

$$\dot{z} = \lambda\, \mathcal{J} H_z \tag{2.6}$$

$$\dot{z} = \lambda\, \mathcal{J} H_{Kz} \tag{2.7}$$

respectively. We seek 2π periodic solutions of (2.6) - (2.7). Theorem 1.4 will be obtained with the aid of the analogous result for (2.7):

Theorem 2.8: If H satisfies (H_3) - (H_5), then for any $K > \bar{r}$ and any $\tau > 0$, (2.7) possesses a nonconstant 2π periodic solution z_K with

$$(2.9) \qquad \int_0^{2\pi} (z_K(t), H_{Kz}(z_K(t)))_{\mathbb{R}^{2n}} dt \le M_1$$

where M_1 is independent of K.

Proof of Theorem 1.4: For each $K > \bar{r}$, by Theorem 2.8 we have a nonconstant solution z_K of (2.7). It suffices to show that for K sufficiently large, $\|z_K\|_{L^\infty} \le K$. Then $H_{Kz}(z_K) = H(z_K)$ so z_K satisfies (2.6). The following lemma provides the desired bound on z_K.

Lemma 2.10: There exists a constant M_2 independent of K such that $\|z_K\|_{L^\infty} \le M_2$.

Proof: By (H_5),

$$(2.11) \qquad H_K(\zeta) \le \hat{\theta}(\zeta, H_{Kz}(\zeta))_{\mathbb{R}^{2n}} + M_3$$

for all $\zeta \in \mathbb{R}^{2n}$ with M_3 independent of K. Taking $\zeta = z_K(t)$ in (2.11), integrating, and using (2.9) yields

$$(2.12) \qquad \int_0^{2\pi} H_K(z_K)dt \le \hat{\theta}\, M_1 + 2\pi M_3 \quad .$$

Since z_K satisfies the Hamiltonian system (2.7), $H_K(z_K)$ is independent of t. Consequently by (2.12),

$$(2.13) \qquad H_K(z_K) \le \frac{\hat{\theta}}{2\pi} M_1 + M_3 \quad ,$$

and the lemma now follows from (2.4) and (2.13).

The proof of Theorem 2.8 will now be carried out using the program sketched in the Introduction. To begin, set

$$(2.14) \qquad I(z) = \int_0^{2\pi} [(p,q)_{\mathbb{R}^n} - \lambda H_K(z)]dt$$

where $z(t) = (p(t), q(t))$. Then $I(z)$ is defined on E, the Hilbert space of $2n$ - tuples of 2π periodic functions which are square integrable and have square integrable first derivatives, i.e. $E = (W^{1,2}(S^1))^{2n}$ under the associated inner product. Formally a critical point of I in E is a weak solution of (2.7).

The first step in our solution procedure is to approximate I on E by a finite dimensional problem. This is easily done here. Let e_k, $1 \le k \le 2n$ denote the usual orthonormal bases in $\mathbb{R}^{2n}$, i.e. $e_1 = (1, 0,..,0)$, etc. Set

$$E_m = \text{span}\{(\sin jt)e_k, (\cos jt)e_k | 0 \le j \le m, 1 \le k \le 2n\}$$

Now we simply consider I restricted to E_m.

The next step in our program is to obtain a nontrivial critical point for $I|_{E_m}$. The following lemma supplies an existence tool. Let $B_r = \{\xi \in \mathbb{R}^j \,|\, |\xi| < r\}$. For $k < j$, let $\mathbb{R}^k = \{\xi \in \mathbb{R}^j \,|\, \xi = (\xi_1,\cdots,\xi_k, 0,\cdots,0\}$ and $(\mathbb{R}^k)^\perp = \{\xi \in \mathbb{R}^j \,|\, \xi = (0,\cdots,0,\xi_{k+1},\cdots,\xi_j\}$.

Lemma 2.15: Let $\Phi \in C^1(\mathbb{R}^j, \mathbb{R})$, $k < j$, and $\Psi:\mathbb{R}^j \to \mathbb{R}$ such that $\Phi(\xi) \le \Psi(\xi)$ for all $\xi \in \mathbb{R}^j$. Suppose

(Φ_1) $\Psi \le 0$ for all $\xi \in \mathbb{R}^k$

(Φ_2) There is a constant $\delta > 0$ such that $\Phi > 0$ in $(B_\delta \setminus \{0\}) \cap (\mathbb{R}^k)^\perp$

(Φ_3) There is a constant $\mu > 0$ such that $\Psi < 0$ in $\mathbb{R}^j \setminus B_\mu$.

Then Φ has a positive critical value b characterized by

$$b = \inf_{h \in \Gamma} \max_{\xi \in \overline{B}_\mu \cap \mathbb{R}^{k+1}} \Phi(h(\xi)) \tag{2.16}$$

where

$$\Gamma = \{h \in C(\overline{B}_\mu \cap \mathbb{R}^{k+1}, \mathbb{R}^j) \,|\, h(\xi) = \xi \text{ if } \Psi(\xi) \le 0\} .$$

Proof: A proof of Lemma 2.15 can be found in [2] or [16].

To apply Lemma 2.15 to $I|_{E_m}$, identify E_m (under $\|\cdot\|_{L^2}$) with $\mathbb{R}^j$ where $j = 2n(2m + 1)$ and take $\Phi = \Psi = I|_{E_m}$. To verify the hypotheses of the lemma, we introduce a convenient bases in E_m. Set

$$\varphi_{jk} = (\sin jt)e_k - (\cos jt)e_{k+n}, 0 \leq j \leq m,\ 1 \leq k \leq n$$
$$\psi_{jk} = (\cos jt)e_k + (\sin jt)e_{k+n}$$
$$\theta_{jk} = (\sin jt)e_k + (\cos jt)e_{k+n}$$
$$\zeta_{jk} + (\cos jt)e_k - (\sin jt)e_{k+n}$$

and take $E^+ = \text{span}\{\varphi_{jk}, \psi_{jk} | j \in \mathbb{N},\ 1 \leq k \leq n\}$, $E^- = \text{span}\{\theta_{jk}, \zeta_{jk} | j \in \mathbb{N},\ 1 \leq k \leq n\}$, $E_m^{\pm} = E^{\pm} \cap E_m$, and $E^0 = \text{span}\{\varphi_{0k}, \psi_{0k} |\ 1 \leq k \leq n\}$. Then E_m^+, E_m^-, E^0 are orthogonal subspaces of E_m whose span is E_m. Let

$$A(z) = \int_0^{2\pi} (p, \dot{q})_{\mathbb{R}^n}\, dt \quad ,$$

the action integral. It is easy to verify that $A > 0$ on $E_m^+ \backslash \{0\}$, $A < 0$ on $E_m^- \backslash \{0\}$, and $A = 0$ on E^0. Choosing $\mathbb{R}^k = E^0 \oplus E_m^-$, $(\mathbb{R}^k)^{\perp} = E_m^+$, and $\mathbb{R}^{k+1} = E^0 \oplus E_m^- \oplus$ span $\{\varphi_{11}\} \equiv V_m$, it now follows from (H_3), (H_4), and (H_5) respectively that (Φ_1), (Φ_2), and (Φ_3) are satisfied. Thus by Lemma 2.15, $I|_{E_m}$ has a positive critical value b_m with corresponding critical point z_m.

The third step in our procedure is to use the minimax characterization of b_m to obtain bounds on b_m and z_m.

<u>Lemma 2.17</u>: There are constants M_4, M_5 independent of m and K and constants M_6, M_7 independent of m such that for all $m > 1$,

(2.18) $$b_m \leq M_4$$

(2.19) $$\int_0^{2\pi} (z_m, H_{Kz}(z_m))_{\mathbb{R}^{2n}}\, dt \leq M_5$$

(2.20) $$\|z_m\|_{L^4} \leq M_6$$

(2.21) $$\|z_m\|_E = (\|\dot{z}_m\|^2_{L^2} + \|z_m\|^2_{L^2})^{1/2} \leq M_7 \quad .$$

Proof: Observe that $h(z) \equiv z \in \Gamma$. Hence by (2.16),

(2.22) $$0 < b_m \leq \max_{B_\mu \cap V_m} I \leq \max_{V_m} I$$

where by (Φ_3) max rather than sup can be used in the right hand inequality. Any function $z \in V_m$ can be expressed as

$$(2.23) \quad z(t) = \|z\|_{L^2}(\zeta(t)\cos\omega + (2\pi)^{-1/2}\varphi_{11}(t)\sin\omega)$$

where $\zeta \in E_m^0 \oplus E^-$, $\|\zeta\|_{L^2} = 1$, and $\omega \in [0,2\pi]$. Choosing $z = \hat{z} \in V_m$ which maximizes $I|_{V_m}$, (2.22) - (2.23) show that

$$(2.24) \quad \lambda \int_0^{2\pi} H_K(\hat{z})\,dt \le \frac{1}{2}\|\hat{z}\|_{L^2}^2 .$$

Using (2.4) and the Hölder inequality to estimate the right hand side of (2.24) yields

$$(2.25) \quad a_3\|\hat{z}\|_{L^2}^{\hat{\theta}^{-1}} - a_4 \le \frac{1}{2}\|\hat{z}\|_{L^2}^2$$

for some constants a_3, a_4 independent of m and K. Since $\hat{\theta} < \frac{1}{2}$, (2.25) provides a bound on $\|\hat{z}\|_{L^2}$ independent of m and K, say

$$\|\hat{z}\|_{L^2} \le M_8 .$$

Returning to (2.22), we find

$$(2.26) \quad b_m \le I(\hat{z}) \le \frac{1}{2} M_8^2 \equiv M_4 .$$

To verify (2.19), note first that since $z_m \equiv (p_m, q_m)$ is a critical point of $I|_{E_m}$)

$$(2.27) \quad 0 = I'(z_m)\zeta = \int_0^{2\pi} [(p_m,\dot{\psi})_{\mathbb{R}^n} + (\varphi,\dot{q}_m)_{\mathbb{R}^n} - \lambda(\zeta,H_{Kz}(z_m))_{\mathbb{R}^{2n}}]dt$$

for all $\zeta = (\varphi,\psi) \in E_m$ where $I'(\xi)\zeta$ denotes the Frechet derivative of I evaluated at ξ and acting on ζ. Using (2.2), (H_5), and some simple estimates, (2.27) with $\zeta = z_m$ gives

$$(2.28) \quad b_m = I(z_m) - \frac{1}{2} I'(z_m)z_m \ge \alpha\int_0^{2\pi} (z_m,H_{Kz}(z_m))_{\mathbb{R}^{2n}}\,dt - a_5$$

where $\alpha = \min(\frac{1}{2} - \theta, \frac{1}{4})$ and a_5 is independent of m and K. Thus (2.19) follows from (2.28) and (2.18).

The definition of H_K and (2.19) yield (2.20).

Lastly (2.27) is employed again with $\zeta = \mathcal{J}\dot{z}_m$ to obtain (2.21). By the Schwarz inequality,

$$\|\dot{z}_m\|_{L^2} \le \lambda\|H_{Kz}(z_m)\|_{L^2} \le a_6(1 + \|z_m\|^3_{L^6}) \tag{2.29}$$

where a_6 depends on K but not on m. Hence

$$\|z_m\|_E \le \|z_m\|_{L^2} + a_6(1 + \|z_m\|^3_{L^6}) . \tag{2.30}$$

The Gagliardo-Nirenberg inequality [17] implies that

$$\|z\|_{L^6} \le a_7\|z\|_E^{1/9} \; \|z\|_{L^4}^{8/9} \tag{2.31}$$

for all $z \in E$. Hence combining (2.30) - (2.31) and (2.20) gives (2.21).

The fourth step in our proof is to use these estimates to get a solution of (2.7). Indeed it now follows from (2.21), the Sobolev Imbedding Theorem [17], and (2.27) that a subsequence of z_m converges weakly in E and strongly in L^∞ to a continuous function $z_K \equiv (p_K, q_K)$ satisfying

$$0 = \int_0^{2\pi} [(p_K, \dot{\psi})_{\mathbb{R}^n} + (\varphi, \dot{q}_K)_{\mathbb{R}^n} - \lambda(\zeta, H_{Kz}(z_K))_{\mathbb{R}^{2n}}]dt \tag{2.32}$$

for all $\zeta = (\varphi,\psi) \in \bigcup_{m>1} E_m \equiv \tilde{E}$. Thus z_K is a weak solution of (2.7). Since $\tilde{E}$ is dense in E, (2.32) implies (2.7) holds a.e. But since $H_{Kz}(z_K)$ is continuous, so is $\dot{z}_K$ and z_K is a classical solution of (2.7). Note also (2.19) implies that z_K satisfies (2.9).

The final step in the proof of Theorem 2.8 is to show that z_K is not a constant. The convergence already established for z_m implies that $b_m = I(z_m) \to I(z_K) \equiv b_k$ along some subsequence. Since $b_m > 0$, $b_K \ge 0$. If z_K is a constant, by (H_3),

$$I(z_K) = -\lambda\int_0^{2\pi} H_K(z_K)dt \le 0$$

so $b_K = 0$. The following lemma shows this is not possible and consequently z_K is nonconstant.

Lemma 2.33: $b_K > 0$.

Proof: A lower bound will be obtained for b_K in terms of a comparison problem. By (H_4) and the definition of H_K, for any $\varepsilon > 0$, there is a constant $A_\varepsilon > 0$ and depending on K such that

$$H_K(z) \le \frac{\varepsilon}{2}|z|^2 + \frac{A_\varepsilon}{4}\sum_{i=1}^{2n} z_i^4 \equiv G(z) \tag{2.34}$$

for all $z \in \mathbb{R}^{2n}$. Set

$$J(z) = \int_0^{2\pi}[(p,\dot{q})_{\mathbb{R}^n} - \lambda G(z)]dt \quad . \tag{2.35}$$

Then by (2.34)-(2.35), $I(z) \le J(z)$ for all $z \in E$. Taking $\Phi = J|_{E_m}$, and $\Psi = I|_{E_m}$, the form of G implies that hypotheses (Φ_1) and (Φ_3) of Lemma 2.15 are satisfied here. Moreover for e.g. $\varepsilon \le \frac{1}{2}$, the quadratic part of J is positive definite on E_m^+ which implies that $J|_{E_m}$ also satisfies (Φ_2). Hence (2.16) defines a critical value c_m of J such that

$$0 < c_m \le b_m \quad . \tag{2.36}$$

If w_m is a critical point of $J|_{E_m}$ corresponding to c_m, then

$$\begin{aligned} c_m &= J(w_m) - \frac{1}{2}J'(w_m)w_m \\ &= \frac{\lambda}{4}A_\varepsilon\int_0^{2\pi}\left(\sum_{i=1}^{2n} w_{mi}^4\right)dt \quad . \end{aligned} \tag{2.37}$$

The estimates of Lemma 2.17 and convergence arguments following it apply to w_m. Hence along a subsequence, $w_m \to w$ satisfying

$$\dot{w} = \lambda \mathcal{J}\, G_z(w)$$

and $c_m \to J(w) \equiv c \ge 0$.

If $b_K = 0$, by (2.36), $c = 0$, and by (2.37), $w \equiv 0$. Therefore $w_m \to 0$ in L^∞. We will show that in fact there is an $\alpha > 0$ such that $\|w_m\|_{L^\infty} \geq \alpha$. Dropping subscripts, we set $w = \bar{w} + W$ where $\bar{w} \in E^0$ and $W \in E_m^+ \oplus E_m^-$. From (2.27) for J' with $\zeta = \bar{w}$, we have

$$(2.38) \qquad 2\pi(\varepsilon|\bar{w}|^2 + A_\varepsilon \sum_{i=1}^{2n} \bar{w}_i^4) = A_\varepsilon \sum_{i=1}^{2n} \int_0^{2\pi} (\bar{w}_i^3 - w_i^3)\bar{w}_i dt \quad .$$

Hence

$$(2.39) \qquad 2\pi \sum_{i=1}^{2n} \bar{w}_i^4 \leq - \sum_{i=1}^{2n} \int_0^{2\pi} (3\bar{w}_i^2 W_i + 3\bar{w}_i W_i^2 + W_i^3)\bar{w}_i dt$$

which together with the Hölder inequality and some simple estimates leads to

$$(2.40) \qquad |\bar{w}| \leq a_8 \|W\|_{L^\infty} \quad .$$

Another application of (2.27) for J' with $\zeta = \mathcal{J}\dot{W}$ yields

$$(2.41) \qquad \|\dot{W}\|_{L^2}^2 \leq 2\varepsilon \|W\|_{L^2}^2 + a_9 \|w\|_{L^\infty}^6$$

where a_9 depends on ε. Since W has mean value zero, it is easy to show that

$$(2.42) \qquad \|W\|_{L^\infty} \leq (2\pi)^{1/2} \|\dot{W}\|_{L^2} \quad .$$

Combining (2.40) - (2.42) gives

$$(2.43) \qquad \|W\|_{L^\infty}^2 \leq 8\pi^2 \varepsilon \|W\|_{L^\infty}^2 + 2\pi a_9 (1 + a_8)^6 \|W\|_{L^\infty}^6 \quad .$$

Since (2.43) is valid for all $\varepsilon \in (0,\frac{1}{2}]$, we choose $\varepsilon = (16\pi^2)^{-1}$. Then (2.43) provides a positive lower bound for $\|W_m\|_{L^\infty}$ and therefore for $\|w_m\|_{L^\infty}$ independently of m.

This completes the proof of Lemma 2.33 and of Theorem 2.8.

Remark 2.44: If H depends explicitly on t in a time periodic fashion and satisfies (H_3) - (H_5), the argument of Theorem 2.8 gives a nonconstant periodic solution of

$$\dot{z} = \lambda \mathcal{J} H_{Kz}(t,z) \tag{2.45}$$

where $H_K(t,z)$ is defined in a similar fashion to (2.1). However the argument of Lemma 2.10 no longer suffices to eliminate the K dependence and some further hypotheses on H seem necessary. See [3].

§3. Proof of Theorem 1.3

In this section we will give an elementary proof of Theorem 1.3 based on Theorem 1.4. To begin we replace H by a more tractable Hamiltonian. The following lemma provides a class of admissable replacements.

Lemma 3.1: Let $H, \overline{H} \in C^1(\mathbb{R}^{2n}, \mathbb{R})$ with $H^{-1}(1) = \overline{H}^{-1}(1)$ and $H_z, \overline{H}_z \neq 0$ on $H^{-1}(1)$. If $\zeta(t)$ satisfies

$$\dot{\zeta} = \mathcal{J} \overline{H}_z(\zeta) \tag{3.2}$$

and $\zeta(0) \in H^{-1}(1)$, then there is a reparametrization $z(t)$ of $\zeta(t)$ which satisfies (1.2). In particular if $\zeta(t)$ is periodic, so is $z(t)$.

Proof: Since $H^{-1}(1), \overline{H}^{-1}(1)$ are level sets for $H, \overline{H}$ respectively, and $\overline{H}^{-1}(1) = H^{-1}(1)$, $H_z(z) = \nu(z)\overline{H}_z(z)$ for all $z \in H^{-1}(1)$ where $0 < \nu(z) \in C(H^{-1}(1), \mathbb{R})$. Moreover since (3.2) is a Hamiltonian system and $\zeta(0) \in H^{-1}(1)$, $\zeta(t)$ lies on $H^{-1}(1)$ for all $t \in \mathbb{R}$. Setting $z(t) = \zeta(r(t))$ where $r(0) = 0$ and r satisfies

$$\frac{dr}{dt} = \nu(\zeta(r(t))) \quad , \tag{3.3}$$

it follows that z satisfies (1.2).

For the periodic case, a bit more care must be exercised since the right hand side of (3.3) is merely continuous and therefore the initial value problem need not have a unique solution. If ζ is T-periodic, let $\overline{t}$ be the first positive value of t such that $r(t) = T$. Replace r by

$s(t) = r(t)$, $t \in [0,\bar{t}]$ and $s(t) = jT + r(t-j\bar{t})$ for $t \in [j\bar{t},(j+1)\bar{t}]$, $j \in \mathbb{Z}$. Then it is easy to verify that $s \in C^1(\mathbb{R},\mathbb{R})$ and $z(t) = \zeta(s(t))$ has period $\bar{t}$.

Proof of Theorem 1.3: It suffices to find a periodic solution of (3.2) for an appropriate choice of $\bar{H}$. As in §2, after a change of time variable, (3.2) becomes

$$\dot{z} = \lambda \mathcal{J} \bar{H}_z \tag{3.4}$$

and we seek a 2π periodic solution of (3.4). To define $\bar{H}$, let $\beta \in C^1(H^{-1}(1), S^{2n-1})$ be the mapping given by (H_2). For each $z \in \mathbb{R}^{2n}\setminus\{0\}$, there is a unique $\alpha(z) \in \mathbb{R}^+$ and $w(z) \in H^{-1}(1)$ such that $z = \alpha w$. Indeed $w(z) = \beta^{-1}(\frac{z}{|z|})$ and $\alpha(z) = |z|\ |w(z)|^{-1}$. Let $\bar{H}(0) = 0$ and $\bar{H}(z) = \alpha(z)^4$, $z \neq 0$. Then $\bar{H} \in C^1(\mathbb{R}^{2n},\mathbb{R})$ and satisfies (H_3) - (H_5). In particular by the homogeneity of $\bar{H}$, $\theta = \frac{1}{4}$ in (H_5) and $\bar{H}_z \neq 0$ if $z \neq 0$. Hence by Theorem 1.4 with $\tau = 2\pi$, (3.4) (with $\lambda = 1$) possesses a nonconstant 2π-periodic solution $u(t)$. Since (3.4) is a Hamiltonian system, $\bar{H}(u(t)) \equiv \rho$, a constant. It need not be the case that $\rho = 1$. However, by the homogeneity of $\bar{H}$, for any $\gamma \neq 0$,

$$(\gamma \dot{u}) = \gamma^{-2} \mathcal{J} \bar{H}_z(\gamma u) \tag{3.5}$$

and

$$\bar{H}(\gamma u) = \gamma^4 \rho \quad . \tag{3.6}$$

Choosing $\gamma = \rho^{-1/4}$, $\bar{H}(\gamma u) \equiv 1$ and γu is a 2π periodic solution of (3.4) with $\lambda = \gamma^{-2}$. The proof is complete.

Remark 3.7: Using the proof of Theorem 2.8, it is not difficult to obtain upper and lower bounds for λ and then via Lemma 3.1 for the period of the solution of (1.2).

§4. Proof of Theorem 1.5

We follow the procedure used in §2, modifying it where necessary. In particular by eliminating hypotheses (H_3) - (H_4), Lemma 2.15 which provided the existence basis for Theorem 1.4 is no longer applicable and a replacement is needed. To get one, we exploit a group structure inherent in our problem which has not yet been employed.

Let $z(t) \in E$. We can write

$$z(t) = \sum_{j=-\infty}^{\infty} \alpha_j e^{ijt} \equiv \varphi(e^{it})$$

where $\alpha_j \in \mathbb{C}^n$ and $\alpha_{-j} = \bar{\alpha}_j$. Let $(L_\sigma z)(t) = z(t+\sigma)$ for $\sigma \in [0,2\pi]$. This family of translations induces an S^1 action on E given by $(\omega\varphi)(e^{it}) = \varphi(\omega e^{it})$ for $\omega \in S^1$. We call mappings of E to E which commute with this action or real valued functions on E which are constant along orbits of the action equivariant maps and subsets V of E for which $L_\sigma : V \to V$ for all $\sigma \in [0,2\pi]$ are called invariant. It is easy to verify that E_m, $E_m^{\pm}$, and E^0 are invariant subspaces of E and $I(z)$ as defined in (2.14) is an equivariant map. Note also that E^0 is a fixed point set for $\{L_\sigma | \sigma \in [0,2\pi]\}$ and there are isotropy subgroups of the action of arbitrary order in S^1.

To take advantage of the above S^1 action, we will use a cohomology index theory developed in [15]. Let $\mathcal{E}$ denote the family of invariant subsets of $E\setminus\{0\}$.

Lemma 4.1: There is a mapping $i: \mathcal{E} \to \mathbb{N} \cup \{\infty\}$, i.e. an index theory such that for all $U, V \in \mathcal{E}$,

1° If there is an $f \in C(U,V)$ where f is equivariant, then $i(U) \le i(V)$.

2° $i(U \cup V) \le i(U) + i(V)$

3° If U is closed, then there is a closed invariant neighborhood V of U such that $i(V) = i(U)$.

4° For $z \in E\setminus E^0$, if $S^1 z = \{L_\sigma z | \sigma \in [0,2\pi]\}$, then $i(S^1 z) = 1$.

5° If F is an invariant subspace of $(E^0)^\perp$, the L^2 orthogonal complement of E^0, then $i(F \cap \mathcal{S}) = \frac{1}{2} \dim F$ where $\mathcal{S}$ is the unit sphere in E.

6° If U is contained in a finite dimensional subspace of E, $I(U) < \infty$ if and only if $U \cap E^0 = \phi$.

Proof: The definition of index and proofs of its properties can be found in [13].

One further property of $i(\cdot)$ will be useful later.

Lemma 4.2: If $F \subset E_m$ is an invariant subspace containing E^0, $\dim F \geq 2n(m+k+1)$, and $U \in \mathcal{E}$ with $U \subset E_m$ and $i(U) \geq n(m-k)+1$, then $F \cap U \neq \phi$.

Proof: The invariance of F implies the same for $F^\perp \cap E_m$. Suppose $F \cap U = \phi$. Then P_m, the L^2 orthogonal projector of E_m to $F^\perp \cap E_m$, belongs to $C(U,(F^\perp \cap E_m)\backslash\{0\})$ and Pm is equivariant. Hence by 1° of Lemma 4.1,

$$(4.3) \qquad i(U) \leq i(P_m(U)) \leq i(U_m) \leq i(\mathcal{S} \cap F^\perp \cap E_m)$$

where U_m denotes the radial projection of $P_m(U)$ to $\mathcal{S} \cap F^\perp \cap E_m$. Since $\dim E_m = 2n(2m+1)$ and $\dim F \geq 2n(m+k+1)$, $\dim F^\perp \cap E_m \leq 2n(m-k)$. Therefore by 5° of Lemma 4.1,

$$(4.4) \qquad i(\mathcal{S} \cap F^\perp \cap E_m) \leq n(m-k) \quad .$$

But (4.3) - (4.4) are contrary to the hypothesis on $i(U)$. Hence $F \cap U \neq \phi$.

Now we can give a variant of Theorem 2.8.

Theorem 4.5: If H satisfies (H_5), then for any $K > \bar{r}$ and $\tau > 0$, (2.7) possesses a 2π periodic solution.

Remark 4.6: As in Theorem 2.8, it is not just existence but also K independent estimates for the solution that are crucial for the sequel. It is inconvenient to present them at this point and they will be stated in the course of the proof.

The notation of §2 will be used in what follows. As earlier we begin by considering $I|_{E_m}$. With the aid of the above index theory and several ideas from [18], we will obtain a family of critical values of this function. The definition of H_K implies that there are constants M and A_M, the latter depending on K, such that

$$(4.7) \qquad H_K(z) \leq M + A_M|z|^4 \equiv \mathcal{H}(z)$$

for all $z \in \mathbb{R}^{2n}$. Since H_K satisfies (H_5), there is an $\bar{R}$ depending on m and K such that for all $R > \bar{R}$, $I(z) < -2\pi\lambda M$ for $z \in E_m\backslash B_R$. We choose any such R for now and will subject it to one further restriction later. Set

$$V_{mk} = E^0 \oplus E_m^- \oplus \text{span}\{\varphi_{ij},\psi_{ij} \mid 1 \leq i \leq k,\ 1 \leq j \leq n\} \quad .$$

Then V_{mk} is an invariant subspace of E_m. Let

(4.8) $\Gamma_m = \{h \in C(E_m, E_m) \mid h$ is an equivariant homeomorphism of E_m onto E_m and $h(u) = u$ if $I(u) \le -2\pi\lambda M\}$.

The reason for normalizing h by the $-2\pi\lambda M$ term will become clearer later. Now define

(4.9) $$c_{mk} = \inf_{h \in \Gamma_m} \max_{u \in \overline{B}_R \cap V_{mk}} I(h(u)) \qquad 1 \le k \le m \ .$$

Lemma 4.10: For any $k \le m$, c_{mk} is a critical value of $I|_{E_m}$ and

(4.11) $$c_{mk} > -2\pi\lambda M \ .$$

We postpone the proof of Lemma 4.10 for now and complete the

Proof of Theorem 4.5: Since $h(z) \equiv z \in \Gamma_m$, by (4.9), (4.11), and (H_5),

$$-2\pi\lambda M < c_{mk} \le \max_{z \in V_{mk}} I(z) \ .$$

Replacing (2.23) by

$$z(t) = \|z\|_{L^2}(\zeta(t) \cos \omega + \xi(t) \sin \omega)$$

where ζ and ω are as earlier and $\xi \in \text{span}\,\{\varphi_{ij}, \varphi_{ij} \mid 1 \le i \le k,\ 1 \le j \le n\}$ with $\|\xi\|_{L^2} = 1$, the proof of Lemma 2.11 proceeds essentially unchanged with the factor of $\frac{1}{2}$ in (2.24) - (2.26) replaced by k. Thus we obtain estimates for c_{mk} and w_{mk}, a corresponding critical point, which are independent of m:

(4.12) $$c_{mk} \le M_4$$

(4.13) $$\int_0^{2\pi} (w_{mk}, H_{Kz}(w_{mk}))_{\mathbb{R}^{2n}}\, dt \le M_5$$

(4.14) $$\|w_{mk}\|_{L^4} \le M_6$$

(4.15) $$\|w_{mk}\|_E \le M_7$$

where $M_4 - M_5$ depend only on k and $M_6 - M_7$ depend on k and K. Now as in §2, a subsequence of w_{mk} converges to a function w_k as $m \to \infty$ and w_k satisfies (2.7).

Thus Theorem 4.5 is proved modulo Lemma 4.10. Once (4.11) has been established, the lemma is a consequence of the following result. For $c \in \mathbb{R}$, let $\mathcal{A}_c = \{z \in E_m | \Phi(z) \leq c\}$ and $\mathcal{K}_c = \{z \in E_m | \Phi(z) = c$ and $\Phi'(z) = 0\}$.

Lemma 4.16: Suppose $\Phi \in C^2(E_m, \mathbb{R})$ is equivariant, there exists an $R > 0$ such that $\Phi(z) < -2\pi\lambda M$ for $z \in E_m \setminus B_R$, $\bar{\varepsilon} > 0$, $c > -2\pi\lambda M$, and $\mathcal{O}$ is any invariant neighborhood of $\mathcal{K}_c$. Then there is an $\varepsilon \in (0, \bar{\varepsilon})$ and $\eta \in C([0,1] \times E_m, E_m)$ such that

1° $\eta(s, \cdot)$ is equivariant for each $s \in [0,1]$

2° $\eta(s, \cdot)$ is a homeomorphism of E_m onto E_m for each $s \in [0,1]$

3° $\eta(s,z) = z$ if $\Phi(z) \notin [c-\bar{\varepsilon}, c+\bar{\varepsilon}]$

4° $\eta(1, \mathcal{A}_{c+\varepsilon} \setminus \mathcal{O}) \subset \mathcal{A}_{c-\varepsilon}$

5° If $\mathcal{K}_c = \phi$, $\eta(1, \mathcal{A}_{c+\varepsilon}) \subset \mathcal{A}_{c-\varepsilon}$.

Proof: With the exception of 1°, the Lemma is a special case of a standard result. Therefore we will only sketch the proof indicating in the process why 1° is satisfied. More details can be found in [19] or [20].

By making $\bar{\varepsilon}$ smaller if necessary, we can assume

$$\bar{\varepsilon} < (c + 2\pi\lambda M) 4^{-1} \equiv \mu \quad . \tag{4.17}$$

The assumption on R implies $\Phi^{-1}([c-\bar{\varepsilon}, c+\bar{\varepsilon}]) \subset \bar{B}_R$ which is compact (and therefore Φ trivially satisfies the Palais-Smale condition in $\bar{B}_R$). Choose any $\varepsilon \in (0, \bar{\varepsilon})$. The function η is determined as the solution of an ordinary differential equation:

$$\frac{d\eta}{ds} = V(\eta) \quad , \quad \eta(0,z) = z \tag{4.18}$$

for $z \in E_m$. To define V, let $\tilde{A} = \mathcal{A}_{c-\bar{\varepsilon}} \cup (E_m \setminus \mathcal{A}_{c+\bar{\varepsilon}})$ and $\tilde{B} = \mathcal{A}_{c+\varepsilon} \cap (E_m \setminus \mathcal{A}_{c-\varepsilon})$. Note that these sets are invariant and therefore $g(z) = \|z-\tilde{A}\|_{L^2}(\|z-\tilde{A}\|_{L^2} + \|z-\tilde{B}\|_{L^2})^{-1}$ is an equivariant function where $\|z-\tilde{A}\|_{L^2}$ denotes the distance (in E_m) from z to $\tilde{A}$. Observe that $g \equiv 0$ on $\tilde{A}$ and $g \equiv 1$ on $\tilde{B}$. Similarly for δ suitably small - see [19] or [20] to make this precise - we can define a Lipschitz continuous equivariant function f such that $f \equiv 0$ on $\{z \in E_m | \|z-\mathcal{K}_c\|_{L^2} \le \frac{\delta}{8}\}$, $f \equiv 1$ on $\{z \in E_m | \|z-\mathcal{K}_c\|_{L^2} \ge \frac{\delta}{4}\}$ and $0 \le f \le 1$. Next define $\varphi : \mathbb{R}^+ \to \mathbb{R}^+$ by $\varphi(s) = 1$ if $s \in [0,1]$ and $\varphi(s) = s^{-1}$ if $s > 1$. Finally set $V(z) = - f(z)g(z)\varphi(\|\Phi'(z)\|_{L^2})\Phi'(z)$ for $z \in E_m$. Then by construction V is uniformly bounded, locally Lipschitz continuous, and equivariant on E_m. It follows that the solution $\eta(s,z)$ of (4.18) exists for all $s \in \mathbb{R}$ and satisfies 1°. The semi-group property for (4.18) gives 2° and the definition of g implies 3°. Lastly 4° - 5° follow as in [19] or [20].

Assuming (4.11) for now, we give the

Proof of Lemma 4.10: Assume first that $H \in C^2$ and there- $I \in C^2(E_m, \mathbb{R})$ and that in (4.8), I is replaced by any function Ψ where $|\Psi(z) - I(z)| \le \mu$ on E_m, μ being defined in (4.17). If c_{mk} is not a critical value of $I|_{E_m}$, we can invoke Lemma 4.16 with $\bar{\varepsilon} = \mu$, $c = c_{mk}$. and $\mathfrak{G} = \phi$. Choose $h \in \Gamma_m$ such that

$$(4.19) \qquad \max_{u \in \bar{B}_R \cap V_{mk}} I(h(u)) \le c_{mk} + \varepsilon \ .$$

By 1° - 2° of Lemma 4.16, $\eta(1,h)$ is an equivariant homeomorphism of E_m onto E_m. Moreover if $\Psi(z) \le - 2\pi\lambda M$, $h(z) = z$ and 3° of Lemma 4.16 shows $\eta(1,h(z)) = z$ provided that $I(z) \notin [c_{mk} - \bar{\varepsilon}, c_{mk} + \bar{\varepsilon}]$. This is certainly satisfied since by our choice of ε; $\Psi(z) \le - 2\pi\lambda M$ implies that $- 2\pi\lambda M - \bar{\varepsilon} \le I(z) \le - 2\pi\lambda M + \bar{\varepsilon} < c_{mk} - \bar{\varepsilon}$. Hence $\eta(1,h) \in \Gamma_m$. Consequently

$$(4.20) \qquad \max_{u \in \bar{B}_R \cap V_{mk}} I(\eta(1,h(u))) \ge c_{mk} \ .$$

But (4.19) and 5° of Lemma 4.16 imply that

$$\max I(\eta(1,h(u))) \le c_{mk} - \varepsilon$$

contrary to (4.20). Hence c_{mk} is a critical value of $I|_{E_m}$.

Now suppose H is merely C^1. Let H_j denote a sequence of C^2 functions which converge to H on $\overline{B}_{K+1}$ in $\mathbb{R}^{2n}$ uniformly in the C^1 norm. Set

$$H_{Kj}(z) = \chi(|z|)H_j(z) + (1 - \chi(|z|))\rho(K)|z|^4$$

for $z \in \mathbb{R}^{2n}$ and

$$I_j(z) = \int_0^{2\pi} [(p,\dot{q})_{\mathbb{R}^n} - \lambda H_{Kj}(z)]dt$$

for $z \in E$. Then the functions $I_j|_{E_m}$ satisfy the hypotheses of Lemma 4.16 for all j sufficiently large and converge to $I|_{E_m}$ in $\overline{B}_R$ uniformly in C^1. For such large j, define c^j_{mk} by (4.9) with I replaced by I_j but Γ_m depending on I. Then c^j_{mk} exceeds $-2\pi\lambda M$ since H_j is close to H. Hence by the case just treated with $\Psi = I$, c^j_{mk} is a critical value of $I_j|_{E_m}$ with corresponding critical point u^j_{mk}. The definition of R implies $u^j_{mk} \in \overline{B}_R$. Hence the compactness of $\overline{B}_R$ and convergence of $I_j|_{E_m}$ to $I|_{E_m}$ imply that along some subsequence $u^j_{mk} \to w_{mk}$ and $c^j_{mk} = I(u^j_{mk}) \to I(w_{mk})$ with w_{mk} a critical point of $I|_{E_m}$. Moreover $I(w_{mk}) = c_{mk}$ as defined in (4.9).

It remains to prove (4.11). This estimate and more follow from a comparison argument. First we define

$$(4.21) \quad \Gamma^*_{mk} = \{S \in E_m \mid S \text{ is compact, invariant, and } S \cap h(\overline{B}_R \cap V_{mk}) \neq \phi \text{ for all } h \in \Gamma_m\} .$$

<u>Lemma 4.22</u>: $\Gamma^*_{mk} \neq \phi$. Indeed if $S \subset B_R \cap E^+_m$ is compact, invariant, and satisfies $i(S) \geq n(m-k) + 1$, then $S \in \Gamma^*_{mk}$.

Proof: Note first that such sets S exist since $i(S \cap E_m^+) = mn$ via 5° of Lemma 4.1. Let $h \in \Gamma_m$. Since $h(z) = z$ for $z \notin B_R$, $h^{-1}(S) \subset B_R$. Therefore $S \cap h(\overline{B}_R \cap V_{mk}) \neq \phi$ is equivalent to $h^{-1}(S) \cap V_{mk} \neq \phi$. Since h is a homeomorphism, $i(S) = i(h^{-1}(S))$ by 1° of

Lemma 4.1. Moreover $\dim V_{mk} = 2n(m+k+1)$. Hence by Lemma 4.2, $h^{-1}(S) \cap V_{mk} \neq \phi$ and $S \in \Gamma^*_{mk}$.

Another set of numbers can now be defined as follows:

$$(4.23) \qquad c^*_{mk} = \sup_{S \in \Gamma^*_{mk}} \min_{u \in S} I(u) \qquad k < m \ .$$

Lemma 4.24: $c^*_{mk} = c_{mk}$

Proof: For each $S \in \Gamma^*_{mk}$ and each $h \in \Gamma_m$, there exists a $\zeta \in S \cap h(\overline{B}_R \cap V_{mk})$. Therefore

$$\max_{h(\overline{B}_R \cap V_{mk})} I \geq I(\zeta) \geq \min_S I$$

from which it follows that $c^*_{mk} \leq c_{mk}$.

To prove equality, observe that for each $h \in \Gamma_m$, there is a $\zeta_h \in \overline{B}_R \cap V_{mk}$ such that

$$I(h(\zeta_h)) = \max_{u \in \overline{B}_R \cap V_{mk}} I \quad (h(u)) \ .$$

Let $S = \{h(S^1 \zeta_h) h \in \Gamma_m\}$ where the notation of 4° of Lemma 4.1 is being employed. Then by construction, $S \in \Gamma^*_{mk}$ and

$$\min_S I \geq c_{mk}$$

so we have equality.

The definition of c^*_{mk} makes it more amenable to lower bounds than c_{mk}. While it is possible to obtain such bounds directly, it is convenient to introduce one more comparison problem. Recall the definition of $\mathscr{G}(z)$ in (4.7). Set

$$(4.25) \qquad \Phi(z) \equiv \int_0^{2\pi} [(p,\dot{q})_{\mathbb{R}^n} - \lambda \mathscr{G}(z)]dt$$

$$= \int_0^{2\pi} [(p,\dot{q})_{\mathbb{R}^n} - \lambda A_M |z|^4]dt - 2\pi\lambda M \ .$$

Thus is the origin of the mysterious term $-2\pi\lambda M$ in the definition of Γ_m. Equation (4.7) implies that

$$(4.26) \qquad \Phi(z) \leq I(z)$$

for all $z \in E$. Therefore

$$(4.27) \qquad c^*_{mk} \geq b^*_{mk} \equiv \sup_{S \in \Gamma^*_{mk}} \min_{u \in S} \Phi(u) \quad .$$

Thus to prove (4.11), it suffices to find an appropriate lower bound for b^*_{mk}. To do this one final set of preliminaries is needed. Any $z \in E^+$ can be written as

$$z = \sum_{j=1}^{n} \sum_{i=1}^{\infty} \alpha_{ij}\varphi_{ij} + \beta_{ij}\psi_{ij} \quad .$$

Therefore

$$(4.28) \qquad A(z) = \frac{\pi}{2} \sum_{j=1}^{n} \sum_{i=1}^{\infty} j(|\alpha_{ij}|^2 + |\beta_{ij}|^2) \quad .$$

It follows that $A(z)^{1/2}$ is a (Hilbert space) norm on E^+. Indeed the closure Y of E^* with respect to $A(z)^{1/2}$ is a subspace of the fractional Sobolev space $(W^{1/2,2}(S^1))^{2n}$.

<u>Lemma 4.29</u>: For all $z \in Y$ and $r \in [2,\infty)$, there is a constant ω_r depending only on r such that

$$(4.30) \qquad \|z\|_{L^r} \leq \omega_r A(z)^{1/2} \quad ,$$

i.e. Y is continuously imbedded in $(L^r)^{2n}$. Moreover the imbedding is compact.

Proof: The first assertion is a special case of standard results on Fourier series. See e.g. the main theorem on integrals of fractional order in [21]. To prove the compactness, observe that (4.30) and the Schwarz inequality imply

$$(4.31) \qquad \|z\|_{L^r} \leq \omega_r \|z\|_{L^2}^{\frac{1}{r}} A(z)^{\frac{r-1}{2r}} \quad .$$

The standard proof of the Rellich lemma - see e.g. [22, p. 169] - implies that Y is compactly embedded in $(L^2)^{2n}$. Therefore if $z_j \rightharpoonup 0$ in Y ($\rightharpoonup$ denoting weak convergence), $z_j \to 0$ in $\|\cdot\|_{L^2}$ and since z_j is bounded in Y, $z_j \to 0$ in $\|\cdot\|_{L^r}$ via (4.31). Hence the imbedding is compact.

Now let $D_{mk} = \text{span}\ \{\varphi_{ij}, \Psi_{ij} \mid k \leq i \leq m,\ 1 \leq j \leq n\}$ and $D_k = \overline{\bigcup_{m \geq k} D_{mk}}$ where the closure is taken in Y. By Lemma 4.29, we have

(4.32) $$\|z\|_{L^4} \le d_k \; A(z)^{1/2}$$

for all $z \in D_k$ where

$$\omega_4 \ge d_k = \sup\{\|z\|_{L^4} \mid z \in D_k \text{ and } A(z) = 1\} \;.$$

Moreover by compactness assertion of Lemma 4.29, there is $\zeta_k \in D_k$ such that $A(\zeta_k) = 1$ and $\|\zeta_k\|_{L^4} = d_k > 0$.

Lemma 4.33: $d_k \to 0$ monotonically as $k \to \infty$.

Proof: The definition of d_k implies that $d_{k+1} \le d_k$. The definition of D_k implies $\zeta_k \rightharpoonup 0$ in Y and hence $d_k = \|\zeta_k\|_{L^4} \to 0$ by Lemma 4.29.

The proof of (4.11) is now completed by combining Lemma 4.24, (4.27), and the following

Lemma 4.34: $b^*_{mk} > -2\pi\lambda M$

Proof: Let $S_{mk} = \{z \in D_{mk} \mid A(z) = \rho^2\}$. By (4.25) and (4.32) we have

(4.35) $$\Phi(z) \ge \rho^2 - \lambda A_M d_k^4 \rho^4 - 2\pi\lambda M$$

for all $z \in S_{mk}$. Choosing $\rho_k = (2\lambda A_M d_k^4)^{-1/2}$ leads to

(4.36) $$\Phi(z) \ge \frac{1}{2}\rho_k^2 - 2\pi\lambda M.$$

Making $R = R(m,K)$ sufficiently large insures that $S_{mk} \subset B_R$. Since S_{mk} is radially homeomorphic to the unit ball in D_{mk}, $i(S_{mk}) = n(m-k+1) \ge n(m-k) + 1$ by 1° and 5° of Lemma 4.1. Therefore Lemma 4.22 shows $S_{mk} \in \Gamma^*_{mk}$. Lastly (4.27) and (4.36) imply $b_{mk} \ge \frac{1}{2}\rho_k^2 - 2\pi\lambda M$ and the proof is complete.

Now finally we can give the

Proof of Theorem 1.5: Fix k. For this prescribed value of k and all m, by Lemma 4.10, c_{mk} is a critical value of $I|_{E_m}$ with a corresponding critical point w_{mk}. Moreover (4.12) and (4.13) provide estimates for c_{mk} and w_{mk} depending on k but independent of m and K. Hence on passing to a limit in m along an appropriate subsequence we get a solution w_k of (2.7) satisfying

$$I(w_k) \equiv c_k \leq M_4 \tag{4.37}$$

$$\int_0^{2\pi} (w_k, H_{Kz}(w_k))_{\mathbb{R}^{2n}} \, dt \leq M_5 \quad . \tag{4.38}$$

The estimate of Lemma 2.10 then shows $\|w_k\|_{L^\infty} \leq M_9$ with M_9 depending on k but not K. Hence choosing $K \geq M_9$, we can assume w_k satisfies (2.6).

To complete the proof, it suffices to show that for k sufficiently large, $\|w_k\|_{L^\infty} > \hat{r}$. If this is not the case, fix K at e.g. $\hat{r}$. By (2.6), for all $k \in \mathbb{N}$ we then have

$$\|\dot{w}_k\|_{L^\infty} \leq \lambda \|H_z(w_k)\|_{L^\infty} \tag{4.39}$$

and therefore

$$c_k = I(w_k) \leq \overline{M} \tag{4.40}$$

where $\overline{M}$ depends on $\max\{|H(z)|, |H_z(z)| \mid |z| \leq \hat{r}\}$ but not on k or K. Since along an appropriate subsequence $c_{mk} \to c_k$, then for any fixed k and large m,

$$c_{mk} \leq \overline{M} + 1 \quad . \tag{4.41}$$

But by Lemma 4.24, (4.27), and (4.36),

$$c_{mk} \geq \min_{S_{mk}} \Phi \geq \frac{1}{2} \rho_k^2 - 2\pi\lambda M = \frac{1}{4\lambda A_M d_k^4} - 2\pi\lambda M \quad . \tag{4.42}$$

Since A_M depends only on K which is fixed and $d_k \to 0$ as $k \to \infty$ by Lemma 4.33, we can violate (4.41) by choosing k large enough in (4.42). This contradiction completes the proof.

Remark 4.43: It is not difficult to show that

$$I(z) = \int_0^{2\pi} [(p,\dot{q})_{\mathbb{R}^n} - \lambda H_K(z)]dt$$

satisfies the Palais-Smale condition in E or in

$(W^{\frac{1}{2},2}(S^1))^{2n}$. This suggests that a direct infinite dimensional minimax characterization of critical values of I may be possible. The difficulty of course lies in finding an appropriate class of sets to work with.

Remark 4.44: (H_5) implies that for each b sufficiently large, $H^{-1}(b)$ is radially homeomorphic to S^{2n-1} and $H_z \neq 0$ on $H^{-1}(b)$. Therefore by Theorem 1.3, there is a periodic solution of (1.2) on this surface. If one could establish better estimates for its period than we have been able to, this approach may provide a simpler proof of Theorem 1.5 than the one just given.

REFERENCES

[1] Weinstein, A., Normal modes for non-linear Hamiltonian systems, Inv. Math., 20, (1973), 47-57.

[2] Rabinowitz, P. H., Free vibrations for a semilinear wave equation, to appear Comm. Pure Appl. Math.

[3] Rabinowitz, P. H., Periodic solutions of Hamiltonian systems, to appear Comm. Pure Appl. Math.

[4] Seifert, H., Periodische Bewegungen mechanischer Systeme, Math. Z., 51, (1948), 197-216.

[5] Weinstein, A., Periodic orbits for convex Hamiltonian systems, preprint.

[6] Berger, M. S., Nonlinearity and Functional Analysis, Academic Press, New York, 1977.

[7] Gordon, W. B., A theorem on the existence of periodic solutions to Hamiltonian systems with convex potential, J. Diff. Eq., 10, (1971), 324-335.

[8] Clark, D. C., On periodic solutions of autonomous Hamiltonian systems of ordinary differential equations, Proc. A.M.S. 39, (1973), 579-584.

[9] Jacobowitz, H., Periodic solutions of $x'' + f(x,t) = 0$ via the Poincáre-Birkhoff theorem, J. Diff. Eq., 20, (1976), 37-52.

[10] Hartman, P., to appear, Amer. J. Math.

[11] Weinstein, A., Bifurcations and Hamiltons principle, preprint.

[12] Moser, J., Periodic orbits near an equilibrium and a theorem by Alan Weinstein, Comm. Pure Appl. Math. 29, (1976), 727-747.

[13] Bottkol, M., Bifucation of periodic orbits on manifolds and Hamiltonian systems, Thesis, New York University, 1977.

[14] Chow, S. N. and J. Mallet-Paret, Periodic solutions near an equilibrium of a non-positive definite Hamiltonian system, preprint.

[15] Fadell, E. R. and P. H. Rabinowitz, Generalized cohomological index theories for Lie group actions with an application to bifurcation questions for Hamiltonian systems, to appear Inv. Math.

[16] Rabinowitz, P. H., Some critical point theorems and applications to semilinear elliptic partical differential equations, to appear Ann. Scuol. Norm. Sup. Pisa.

[17] Friedman, A., Partial Differential Equations, Holt, Rinehart, and Winston, Inc., New York, 1969.

[18] Ambrosetti, A. and P. H. Rabinowitz, Dual variational methods in critical point theory and applications, J. Functional Anal., 14, (1973), 349-381.

[19] Clark, D. C., A variant of the Ljusternik-Schnirelman theory, Ind. Univ. Math. J., 22, (1972), 65-74.

[20] Rabinowitz, P. H., Variational methods for nonlinear eigenvalue problems, Eigenvalues of Non-linear Problems, Edizioni Cremonese, Rome, 1974, 141-195.

[21] Zygmund, A., Trignometric Series, Cambridge University Press, New York, 1959.

[22] Bers, L., F. John, and M. Schechter, Partial Differential Equations, Interscience Publishers, New York, 1964.

This research was supported in part by the Office of Naval Research under Contract No. N00014-76-C-0300 and by the U.S. Army under Contract No. DAAG29-75-C-0024. Any reproduction in part or in full for the purposes of the U.S. Government is permitted.

Department of Mathematics
University of Wisconsin
Madison, Wisconsin 53706

Index

A
B
C 8
D 9
E 0
F 1
G 2
H 3
I 4
J 5